The Makerspace Workbench

Adam Kemp

ER **MEDIA**™
, CA

The Makerspace Workbench

by Adam Kemp

Printed in Canada.

Published by Maker Media, Inc., 1005 Gravenstein Highway North, Sebastopol, CA 95472.

Maker Media books may be purchased for educational, business, or sales promotional use. Online editions are also available for most titles (*http://my.safaribooksonline.com*). For more information, contact O'Reilly Media's corporate/institutional sales department: 800-998-9938 or *corporate@oreilly.com*.

Editors: Brian Jepson and Michelle Hlubinka
Production Editor: Rachel Steely
Copyeditor: Octal Publishing, Inc.
Proofreader: Kiel Van Horn
Indexer: WordCo Indexing Services, Inc.

Cover Designer: Jason Babler
Interior Designers: Monica Kamsvaag and Nellie McKesson
Illustrators: Adam Kemp and Rebecca Demarest
Photographer: Adam Kemp

September 2013: First Edition
Revision History for the First Edition:

2013-08-27:   First release

See *http://oreilly.com/catalog/errata.csp?isbn=9781449355678* for release details.

ISBN: 978-1-449-35567-8

[TI]

For my boys and to making things together.

Table of Contents

Preface

Welcome to the world of tomorrow! What started off in the basements of hardware hackers and the minds of science fiction authors has blossomed into one of the world's foremost places for innovation and design: the Makerspace. These spaces for creation have emerged as a safe haven for tinkerers, artists, engineers, scientists, and anyone with an inspired spirit to tap into their imaginations and create.

So, where do you draw the line between a run-of-the-mill shop and a Makerspace? Well, the answer isn't easy to define. Makerspaces serve as portals for learning, collaboration, problem solving, and self expression, and rapidly evolve and reconfigure to support the interesting and new. Where a shop strives to support a common set of tools and manufacturing capability, the Makerspace relies on an unknown tomorrow. This care-free approach for establishing such a space allows for universal acceptance. Anyone can play. There is no height restriction.

Why Make?

Simple. Because it is so much fun. Making makes your brain hurt, your fingers sting and your room dirty: things you just can't buy. The Maker movement has brought a philosophy of sharing, acceptance, and creativity and can be applied to the establishment of any Makerspace. The following manifesto encapsulates many of the fundamental understandings that any Makerspace should embrace:

The Makerspace Manifesto

- Everyone is a Maker.
- Our world is what we make it.
- If you can imagine it, you can make it.
- If you can't open it, you don't own it.
- We share what we make and help each other make what we share.
- We see ourselves as more than consumers—we are productive; we are creative.
- Makers ask, "What can I do with what I know?"
- Makers seek out opportunities to learn to do new things, especially through hands-on, do-it-yourself (DIY) interactions.
- The divisions between subjects like math and art and science dissolve when you are making things.
- Making is an interdisciplinary endeavor.
- It's all right if you fail, as long as you use it as an opportunity to learn and to make something better.
- We're not about winners and losers. We're about everyone making things better.

- We help one another do better. Be open, inclusive, encouraging, and generous in spirit.
- We celebrate other Makers—what they make, how they make it, and the enthusiasm and passion that drives them.

Now, rather then letting someone else's closed design serve as the basis for you project, Makerspaces and its Makers are here to help. Ask questions, bend rules, and most important, learn by doing. Whether in your basement, community center, commercial property, or school, every space has the potential to inspire minds and Make great things. So, put on your safety glasses and let's begin the journey into Making your Makerspace!

Moving Through This Book

I created this book to serve as a tool for everyone interested in Making, and it should not be kept on bookshelves. Rather it should be covered in notes, torn, and slightly charred sitting next to your next amazing creation. Each chapter is designed to educate the reader about one aspect of the Makerspace environment through the use of reference text, informative tips, and technology breakdowns. Sections conclude with one of more than 30 topic-reinforcing projects that serve as a means of buttressing that section's topic, producing an artifact that is useful for the Makerspace environment or as just a fun and exciting activity. The book:

- Acts as a starting point and reference for both establishing and Making in a Makerspace by providing in-depth overviews of a Makerspace's tools and technology.
- Provides insight for teachers both inside and outside of the standard shop-type environment and shows them the potential Makerspace technology has in improving curriculum and the classroom experience.
- Illustrates the potential a Makerspace has to inspire in the world of design, invention, and

manufacturing through the use of low-cost, open source hardware and software.

What Makes Up Your Makerspace?

Well, there is a good chance that you already have many of the skills, tools, and equipment that form the foundation of a great makerspace. These things can range from a simple soldering iron to a full-blown workshop—and not all are physical. Makerspaces thrive on the creativity and imagination of the Makers inside, and that is surely something you cannot buy. When you combine the minds of the Makers and provide them with a space and the tools for Making, you set the catalyst for your Makerspace to come to life. So, what should you expect to have in order to begin or continue the evolution of your Makerspace? Let's take a look. Table P-1 provides a quick list of the primary equipment that are commonly found in the Makerspace environment and can give you insight as to where you are in your journey.

Table P-1. Makerspace equipment: from start to finish

	In-Development	Well-Developed	Very Well-Developed	Exceptional
Creativity	X	X	X	X
Computing Technology	X	X	X	X
Soldering Equipment	X	X	X	X
Hand Tools		X	X	X
Power Tools		X	X	X
Standing Power Tools				X
3D Printer			X	X
Laser Engraver				X

During my past eight years of teaching students in a Makerspace-style environment, I have noticed a trend in the types of projects my students create using the unique Makerspace equipment. Table P-2 outlines my "Top 10" Makerspace projects. Where will your Makerspace take you?

Table P-2. Adam's "Top 10" Things to Make in a Makerspace

Rank	Thing	Equipment Used
1	Anything Arduino	Arduino and misc. circuit components
2	Hand-etched circuits	EAGLE, bare PCB, kapton tape, and ferric chloride
3	Children's toys	3D printer or laser engraver
4	Wheeled robots	3D printer or laser engraver
5	Autonomous RC vehicles	Arduino and ArduPilot
6	Model rockets	Laser engraver
7	Home automation and monitoring	RPi, Arduino, and misc. electronic components
8	Project enclosures	3D printer or laser engraver
9	Go-karts battery/gas powered	Lawn mower parts, wheels, metal, and welding gear
10	Engraved cellphone/laptop cases	Laser engraver

Parts and Material Sources

Many of the tools required to begin the design of your Makerspace are provided in this book, and the rest can be obtained easily and inexpensively. One of the neatest things about the construction of the Makerspace is its uniqueness and individuality. The Maker movement has shown, time and again, that your budget should not be a limiting factor. There is an almost endless source of components and hardware that can be found locally for very little money, and sometimes even free. If you have never been to a thrift store or yard sale, you are in for a treat. By looking at, say, an old printer, and with this book's help, you will begin to see the magic behind the curtain. And, for a few dollars can obtain all of the components necessary to make your next invention.

We recommend that you take a look at two other documents we've produced for suggestions, checklists, and images of gadgets, tools, workspaces, and more:

- *Make:* magazine's special issue, the *2011 Ultimate Workshop and Tool Guide*
- High School Makerspace Tools & Materials (*http://makerspace.com/maker-news/makerspace-playbook*)

Electronics

Adafruit / Maker Shed / Robotshop / Seeed Studio / SparkFun
Creators and distributors of the finest Maker technology: *http://www.adafruit.com*, *http://www.makershed.com*, *http://www.robotshop.com*, *http://www.seeedstudio.com*, and *http://www.sparkfun.com*

All-Electronics
Sells "pre-owned" and surplus stuff for super cheap. Its inventory frequently rotates, so stock up when you find something you like! *http://www.allelectronics.com*

Digikey / Mouser
Semiconductor and electronic component distributors. Pay attention to quantity discounts as it is often cheaper to buy more of something rather than the quantity you need. *http://www.digikey.com*, *http://www.mouser.com*

Frys / Microcenter / RadioShack
Carry many tools and components for Makerspace projects. Some even carry Arduino, Maker Shed, Parallax, and SparkFun components! *http://www.frys.com*, *http://www.microcenter.com*, *http://www.radioshack.com*

Tools and Materials

ACMoore / Michaels
Suppliers of tools, raw materials, and craft supplies. Make sure you check their flyer for

decent coupons. *http://www.acmoore.com*, *http://www.michaels.com*

Burden's Surplus Center
Has just about everything you need for larger projects. Including motors, actuators, gearing, and so on. *http://www.surpluscenter.com*

Craigslist
A free marketplace for virtually anything. Be smart when you buy or sell. Make transactions in well-lit and occupied places. I personally like to use the local fastfood restaurant. *http://www.craigslist.org*

Grainger / MCMaster-Carr / MSC
Distributors of hardware, raw materials, tools, and fasteners. *http://www.grainger.com*, *http://www.mcmaster.com*, *http://www.mscdirect.com*

Home Depot / Lowes
Home improvement stores that supply tools, raw materials, fasteners, and so on. *http://www.homedepot.com*, *http://www.lowes.com*

Kelvin
Great supplier for educational kits and materials. *http://www.kelvin.com*

Local Hardware Stores
Search online for small and local hardware stores. These tend to have better selections of small and unique parts when compared to the larger chain stores.

Conventions Used in This Book

The following typographical conventions are used in this book:

Italic
Indicates new terms, URLs, email addresses, filenames, and file extensions.

`Constant width`
Used for program listings, as well as within paragraphs to refer to program elements such as variable or function names, databases, data types, environment variables, statements, and keywords.

`Constant width bold`
Shows commands or other text that should be typed literally by the user.

`Constant width italic`
Shows text that should be replaced with user-supplied values or by values determined by context.

This box signifies a tip, suggestion, general note, or warning.

Using Examples

This book is here to help you get your job done. In general, if this book includes code examples, you may use the code in this book in your programs and documentation. You do not need to contact us for permission unless you're reproducing a significant portion of the code. For example, writing a program that uses several chunks of code from this book does not require permission. Selling or distributing a CD-ROM of examples from MAKE books does require permission. Answering a question by citing this book and quoting examples does not require permission. Incorporating a significant amount of examples from this book into your product's documentation does require permission.

We appreciate, but do not require, attribution. An attribution usually includes the title, author, publisher, and ISBN. For example: "*The Makerspace Workbench* (MAKE). Copyright 2013 Adam Kemp, 978-1-449-35567-8."

If you feel your use of code examples falls outside fair use or the permission given above, feel free to contact us at *bookpermissions@makerme dia.com*.

Safari® Books Online

Safari Books Online is an on-demand digital library that delivers expert content in both book and video form from the world's leading authors in technology and business.

Technology professionals, software developers, web designers, and business and creative professionals use Safari Books Online as their primary resource for research, problem solving, learning, and certification training.

Safari Books Online offers a range of product mixes and pricing programs for organizations, government agencies, and individuals. Subscribers have access to thousands of books, training videos, and prepublication manuscripts in one fully searchable database from publishers like MAKE, O'Reilly Media, Prentice Hall Professional, Addison-Wesley Professional, Microsoft Press, Sams, Que, Peachpit Press, Focal Press, Cisco Press, John Wiley & Sons, Syngress, Morgan Kaufmann, IBM Redbooks, Packt, Adobe Press, FT Press, Apress, Manning, New Riders, McGraw-Hill, Jones & Bartlett, Course Technology, and dozens more. For more information about Safari Books Online, please visit us online.

How to Contact Us

Please address comments and questions concerning this book to the publisher:

Maker Media, Inc.
1005 Gravenstein Highway North
Sebastopol, CA 95472
800-998-9938 (in the United States or Canada)
707-829-0515 (international or local)
707-829-0104 (fax)

We have a web page for this book, where we list errata, examples, and any additional information. You can access this page at *http://oreil.ly/ Make_workbench*.

To comment or ask technical questions about this book, send email to *bookquestions@oreil ly.com*.

Maker Media is devoted entirely to the growing community of resourceful people who believe that if you can imagine it, you can make it. Maker Media encourages the do-it-yourself mentality by providing creative inspiration and instruction.

For more information about our publications, events, and products, see our website at *http:// makermedia.com*.

Find us on Facebook: *https://www.facebook.com/ makemagazine*

Follow us on Twitter: *https://twitter.com/make*

Watch us on YouTube: *http://www.youtube.com/ makemagazine*

Making the Space   1

IN THIS CHAPTER

- Choosing the Location
- Designing the Space
- Safety

In a perfect world, the Makerspace is as common as a coffee shop. You become a member, whether it be paid or free, and when an idea hits you, you know where to go. As you enter the building into a well-lit, clean environment, you see Makers of all ages working tirelessly on their exciting projects. Around the perimeter of the room, you see people on computers with open source CAD software designing circuits and mechanisms, 3D printers extruding parts for more 3D printers, and soldering irons being used to mount components on custom Arduino shields. Near the windows you hear the hum of a squirrel-cage blower removing the smoke from a laser cutter that is meticulously outlining the frame of a quad-rotor helicopter. And in the center of the room, there is a group of students working on their robots for the next FIRST Lego League competition. Paradise.

This dream is quickly becoming reality. Makerspaces of all types are popping up around the world and are opening their doors to the future of Making. For these establishments to become successful at their missions, it is fundamentally important that they start with a good design. This chapter will discuss possible locations for your space, tools for laying it out, and ensuring its occupants have a safe environment to work (Figure 1-1).

Figure 1-1. *My project area.*

Choosing the Location

The first question asked when designing a Makerspace is "Where am I going to put it?" This decision sets the stage for determining the types of equipment, materials, and projects the space can support. It sheds light on just how many people can occupy the space as well as its potential for growth. Choosing the location is mainly dependent on the direction you want the space to go. What kind of projects do you want to support? Are they craft based or do they require sophisticated machinery? This section is designed to assist with this decision and will help develop an understanding of the proposed locale's benefits and drawbacks. Ultimately, choosing the optimal location ensures that its participants can function safely and effectively while they work.

Understanding the Constraints

Although each location possesses unique design constraints, there are common elements that are universal. In understanding each of these elements and the limitations they expose, you will be better able to choose, design, and ultimately construct your Makerspace. These constraints will also help to determine just what kind of equipment your Makerspace needs. Table 1-1 helps to illustrate different methods and tools required for completing common tasks. It might turn out that your space can get away with simple handheld power tools rather than the larger standing type. Or, that there are different ways for making a hole that doesn't involve a drill.

Table 1-1. Different tools for the same task

Task	Tech Level	Tool
Making a hole in <1/4" Wood/ Plastic	Low	Mechanical drill or hole punch
	Medium	Hand Drill
	High	Laser Engraver
Making a hole in >1/4" Wood/ Plastic	Low	Hand drill
	Medium	Drill Press
	High	CNC Router
Making a hole in sheet metal	Low	Sheet Metal Punch
	Medium	Hand drill w/ wood backing block
	High	Pneumatic punch
Making a hole in metal plate	Low	Hand drill w/ hole saw
	Medium	Plasma cutter
	High	CNC Mill
Cutting a profile in wood/ plastic	Low	Hand or coping saw
	Medium	Jig, scroll, or band saw
	High	Laser engraver
Cutting sheet metal	Low	Sheet metal hand shears
	Medium	Floor shear
	High	Pneumatic or electric hand shears

Task	Tech Level	Tool
Cutting metal plate	Low	Hack saw
	Medium	Reciprocating saw
	High	Plasma cutter
Constructing a 3D object	Low	Hand model and cast
	Medium	3D Printer
	High	CNC Mill
Soldering a PCB	Low	Unadjustable soldering iron
	Medium	Adjustable soldering iron
	High	Reflow oven
Making circuit boards	Low	Toner transfer and hand etch
	Medium	Photo transfer and hand etch
	High	PCB Mill

Size

The size of your Makerspace is ultimately the biggest constraint. It dictates how many people can safely work at one time, the types and quantity of equipment you can support, and the size of the projects that can be worked on. A good rule of thumb for determining a number of occupants in your Makerspace is to allocate 50 sq. ft. of space per person: that's a roughly 7 ft × 7 ft area. This allotment allows for safe use of floor space, especially as the occupants will be working in a lab environment. You can find more information in the *BOCA National Building Code/1996*, Building Officials & Code Administrators International, Inc., 1996.

Equipment and technology take up space, require power, and often require some amount of ventilation for proper and safe operation. Table 1-2 is a list of common large equipment found in the Makerspace environment and their approximate size and power requirements.

Table 1-2. Common Makerspace equipment

Type	Size (ft)	Power (Watts)
3D Printer	1 × 1	100
Laser Cutter w/ Ventilation	3 × 5	1500
Standing Drill Press	2 × 3	350
Table-Top Drill Press	1 × 2	125
Standing Band Saw	2 × 3	350
Table-Top Band Saw	1 × 2	120
Soldering Iron	1 × 1	75
Heat Gun	1 × 1	1500
Hot Plate	1 × 1	750

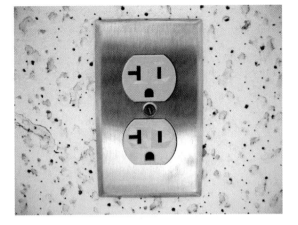

Figure 1-2. NEMA 5-20 outlets feature a horizontal slot for a 20 amp plug.

Power

At the end of the day, someone has to pay the power bill. This constraint is important to understand as many of the pieces of equipment your Makerspace will use require a lot of power. Tools like heat guns and hot plates as well as equipment like laser cutters and their ventilation systems consume hundreds of watts of energy during use.

There are typically two types of outlets that will be available: NEMA 5-15 and NEMA 5-20 (Figure 1-2). Their design dictates how much energy that electrical branch can supply, specifically 15 amps and 20 amps. If you go over the available current, like you would if you used 4 heat guns on one outlet, you run the risk of tripping a circuit breaker, or in the worst case, starting a fire.

Whether you are creating a public or private Makerspace, it is imperative that you follow your local and state rules and regulations pertaining to fire-code and safety. A good place to locate this information is through the National Fire Protection Association at http://www.nfpa.org and the Occupational Safety & Health Administration at http://www.osha.gov. This book should not serve as the only source of information regarding outfitting and occupying a space, and it is your responsibility and discretion to ensure that your Makerspace follows the rules.

An alternative to calculating power consumption is to use an in-line or inductance type power meter (see Figure 1-3). These devices are designed to measure and display immediate power consumption, power consumption with respect to time, current draw, and voltage. They also have the ability to predict the cost in electricity to operate that piece of equipment, which could prove to be very beneficial for understanding the costs involved in operating your Makerspace.

Power Calculation

With DC circuits, we can simply calculate power using P = IV and, conversely, the current by using I = P/V. This equation holds true for instantaneous power in an AC circuit, yet the average power of an AC circuit is determined based on its power factor. You can calculate your equipment's potential current consumption prior to its use by using the following formula:

$$I = W / (PF \times V)$$

I	Current in amps
W	Power in watts
PF	Power factor
V	Voltage in volts

The *power factor* describes the ratio between the power actually used by the circuit (*real power*) and the power supplied to the circuit (*total power*). This value ranges from 0 to 1 and can be difficult to pin down without a good understanding of the internal circuitry or through physical testing. Typically, resistive loads, like heaters and lamps, receive a 1.0 power factor. Equipment containing motors have a power factor less than 1, requiring more power than would be necessary if the circuit were purely resistive, and directly correlates to the efficiency of the system. For the most part, the equipment that you will be using in your Makerspace will not be drawing a large amount of power. Those that do will typically identify their power requirements either on a sticker or within the documentation.

Figure 1-3. *Digital watt meters can accurately display a piece of equipments' power consumption in real time.*

Ventilation

Nobody wants to work in a stinky room and the fumes emitted by Makerspace technology can quickly become a problem. The necessity for proper ventilation poses a serious design constraint if your Makerspace is going to support equipment that produces fumes. Technology like 3D printers, soldering irons, heat guns and plates, and most especially laser cutters are the primary contributors of potentially harmful fumes. Normal room ventilation systems (Figure 1-4) either recirculate the air after it passes through a series of filters or it is pumped in fresh. As the existing ventilation systems are something that cannot easily be changed, localized vapor removal methods need to be implemented.

Figure 1-4. *The ventilation system directs unwanted fumes outside of the building.*

Noise

The fact of the matter is this: tools make noise. On paper this might not seem like a very big issue, but the quantity of noise a machine generates can and will dramatically affect the layout of a Makerspace. This constraint is especially important to consider when implementing Makerspaces in schools and libraries. Even though these Makerspaces might be located in a room separated from the rest of the building, most commercial structures have drop ceilings and false walls. Sound also has the tendency to travel through duct work and will "broadcast" the Makerspace's activities throughout the rest of the building. Table 1-3 illustrates some of the more common, and noisy, Makerspace equipment and just how much noise you can expect them to produce while in operation.

Table 1-3. Equipment noise comparison

Reference	Tool/Equipment	Noise Level (dB) [a]
Rock band		110
	Hammer	100.4
Lawn mower		100
	Reciprocating saw	95.5
	Band saw	91.3
Blender		90
	Hand Drill	89.9
	Hack saw	89.7
City traffic		85
	Laser engraver w/ exhaust	82.2
Vacuum cleaner		75
	Drill press	72.3
Normal conversation		60

[a] Noise level readings were taken approximately 3 ft from the source using a TENMA 72-860 sound-level meter

If your equipment produces a lot of noise, you'll want to get some ear protection (Figure 1-5).

Figure 1-5. *Ear protection should always be worn when working in a loud environment.*

The Library Makerspace

The public establishment of a Makerspace is a marvelous idea. It acts as a common place for our youth to learn and explore engineering concepts, community members to organize and share designs, and it offers an extension to the classroom environment. Libraries happen to fit this bill perfectly. With their endless source of research materials, Internet access, and public atmosphere, what better place for a Makerspace. Wouldn't it be nice to have your public library support the basics for Making? Why shouldn't it?

Most libraries have allocated space that patrons can reserve for nonprofit events that consist of either an isolated room or a specific section of the library's floor space. Optimally, the library has space allocated for long-term installations. Because every library is different, you should check with your local branch's website for more information. Whereas many Makerspaces rely on equipment that involves substantial setup or is inconvenient to move, having a permanent space is incredibly convenient. If it turns out that the library only reserves its space for short periods of time, the nature and direction of your Makerspace will need to be flexible enough to conform accordingly.

After the space has been selected, it is imperative to inform the library's director about the nature of your Makerspace. This is important to do before you set up because the equipment and practices you intend to employ might conflict with the library's rules and regulations. During this conversation it might be beneficial to emphasize the following benefits a Makerspace can bring to the library and community:

1. It aids to excite young minds about engineering and manufacturing.
2. It can serve as an educational outreach tool for local schools.
3. It will act as an instructional environment for the community.

A public library is the ideal location for a Makerspace, though there are some caveats, the first of which being noise. Anyone who has ever been fortunate enough to bring their child into a library understands the magnitude of this problem. Many a time has the lovely librarians at our local library given my son the death stare for, well, being a child. This noise requirement places a pretty big constraint on the various resources the Makerspace can offer. Tools like drill presses and band saws inherently produce a great deal of noise when operating, whereas devices like the 3D printer and soldering irons do not. If you look at the common Makerspace tools and equipment, you can quickly compile a list that meets the environmental constraints.

In addition to the availability of space, each library system is designed to monitor and track its patrons via a unique ID number assigned with the library card. This system could easily be reused by the Makerspace to log user information, record attendance, check-out/in materials and hardware, and so on, making the Makerspace much more consistent and sustainable.

Another constraint imposed by the library environment is the potential lack of a permanent location. This applies to Makerspaces that operate as a scheduled event that reserves one of the library's common spaces. If this is the case, an effort should be made to propose a more permanent location because it saves time and accommodates long-term projects. In the meantime, let's take a look at how you can design a Makerspace that uses a temporary location.

The last thing you want to do is spend significant time setting up and tearing down your Makerspace. But, even in a nonpermanent location, there are some tricks you can employ that will help expedite this process.

Project Boxes

Project boxes, like the one shown in Figure 1-6, are an effective way of organizing designated project areas and reducing the time required for the setup and teardown of work area equipment. The boxes should be designed to support a specific task and contain all of the equipment and tools required. For example, an Adhesives Project Box would contain a glue gun, glue sticks, epoxy, wood glue, super glue, mixing cups and sticks, and minor surface preparation materials. When the adhesives project is complete, the box can be quickly reassembled and ready to support the next task.

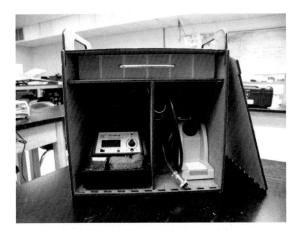

Figure 1-6. *Designated project boxes are a good way to keep work areas clutter free and allow for a quick cleanup.*

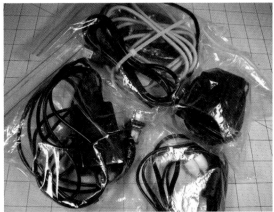

Figure 1-7. *Sealable sandwich bags are great for storing loose wires.*

Taming Wires

Even wires that are properly bundled pose a serious organizational problem. A good method for tackling this problem is to keep each, or similar, wires bundled in sealable sandwich bags (see Figure 1-7). This seemingly simple solution is incredibly effective at preventing the "wire monsters" that occur when bundles of cable are shoved into storage. By quickly coiling similar cables and storing them in sealable bags, they are then easily recovered with little mess. It is amazing how much time gets eaten up when trying to untangle a single cable from a nest of its closest friends.

With a little luck and preparation, the limits imposed by the library Makerspace are minor at best. Remember, the goal of your Makerspace should be to provide the most effective work environment possible, and by understanding the limitations beforehand, more time can be spent working on projects.

The School Makerspace

Workshop environments in our schools are disappearing at an alarming rate. In addition, departments like Technology Education are not considered "core" areas, and the idea exists that workshop environments don't develop skills necessary to go to college. As a teacher, I have witnessed firsthand a surge of interest in problem-based curriculum from both our youth and their parents due to its ability to engage students and help them retain the knowledge. This is why the marriage between the classroom and the Makerspace is so potent; it fills the gap between classroom theory and the physical world.

Historically, sparse classroom budgets have been the root cause for a lack of modern equipment in the classroom. This fact has remained true for years when you consider an entry level 3D printer could cost more than $20,000! Now, a derivative of the technology can be purchased with the proceeds of a single bake sale, or even through parent donations. The beauty of the Makerspace is its ability to not only inspire students but accelerate their knowledge intake through exciting and imaginative curricular application. To facilitate this, schools need to consider the design constraints imposed by Makerspace equipment and how it might affect classroom layout.

There are two ways that a Makerspace can be integrated into your school; either as part of the existing classroom environment or as an entity unto itself. Although each present different challenges, they can be profoundly effective in assisting and inspiring students.

The Makerspace Classroom
As part of the classroom environment, the Makerspace mentality and equipment can be instrumental to the success of the curriculum and engagement of the students. Students embrace the responsibility of using technical equipment, and when they see the potential this equipment provides, their excitement will help motivate their peers.

The decision to merge Makerspace and classroom should extend department wide because it helps to improve knowledge retention among students. This idea of *vertical articulation* applies typically to curriculum yet is just as effective when working with equipment. For example, students are educated and conditioned to use calculators in their math classes. When they leave class and go to a science course that needs to solve an equation, chances are they are going to reach for that calculator. If every classroom had a 3D printer, they would reach for that, too.

The benefits Makerspace technology can afford the classroom environment is astounding. One of the largest hiccups that has prevented innovative exploration in the classroom is the presence of standardized curriculum. With the help of Makerspace technology, innovation and imagination can now supplement and support the standardized curriculum, making the classroom more exciting and engaging for student and teacher alike.

Most classrooms integrate the idea of a "lab" or "activity" that takes the students away from the books and requires them to apply the concept in a physical manner. This allocated time is optimal for the education and integration of Makerspace technology. The following are potential content areas in which Makerspace equipment could benefit the classroom environment:

Science
Makerspace technology can be used to assist in the physical modeling and assembly of chemical compounds, cell and bone structures, and in developing an understanding of how data is acquired and analyzed.

Technology and Engineering
Makerspace technology can supplement a wide array of projects that would normally require multithousand-dollar machines. It directly ties in to courses that focus on architecture, design and manufacturing, robotics, industrial and mechanical engineering, electronics, and virtually all other technology education and engineering curricula.

Art
Makerspace technology can provide a medium for a large number of artistic projects. This can include modeling, photography, computer-controlled art, light and sound, and the list goes on.

Mathematics
Makerspace technology can help illustrate many mathematical concepts through the production of physical objects. Equations and their relationships can be physically constructed, altered, and computed all within the classroom.

The Standalone School Makerspace
In the event that the Makerspace cannot coexist with the classroom environment, it might function better as an entity unto itself. This standalone Makerspace can serve as a "go-to" resource for individual classrooms and student projects alike. Because Makerspace equipment can be implemented into areas with relatively large space constraints, the repurposing of a small classroom or teacher workroom can make the decision a breeze.

With any workshop environment, it is important to clearly define the person in charge of the materials and equipment as well as the space's limitations. This supervisor, whether it be teacher or otherwise, is solely responsible for the space's maintenance and upkeep, because classrooms depend on the resources. As you can see, a situation where multiple classrooms are dependent on a single Makerspace ends up in a scheduling and resource nightmare. Compounded with the need for students to be directly involved with using the equipment, it is important to ensure that this space can in fact support the classrooms that depend on it. It isn't fair to bring 35 students into a small Makerspace and expect them to all get a chance to use the equipment. This is why it is so important to understand the Makerspace's limitations in order for it to succeed.

Aside from operating as a classroom resource, the Makerspace can serve as a resource for individual student projects and initiatives. Depending on your school's schedule, there is often time when students are not required to be in class; for example, during lunch, before or after school, or during designated club/activity time. It can also provide a manufacturing resource for student organizations like TSA, FIRST Lego, FIRST Robotics, and Odyssey of the Mind.

The Garage Makerspace

Garages are not just for parking cars. They can contain the world's most elaborate and ingenious creations and if you happen to be one of those fortunate enough to have a garage, we are about to unlock its potential. With today's wealth of open source technology and the "share and share alike" mentality, setting up an in-home Makerspace is easier than ever. The underlying goal of any Makerspace should be to provide an environment that supports and inspires its members to create with tools that were once not possible. The fact that we have tools like open source 3D printers, laser cutters, electronics, and other resources is the result of the overwhelming generosity of the Maker community. This mantra should be reflected in your Makerspace and afforded to willing members of the community.

The garage Makerspace has incredible potential to support local organizations like the Girl Scouts, Boy Scouts, FIRST Lego and FIRST Robotics clubs. By providing a safe and effective work environment for these youth to operate, you are not only doing the community a great service, but are providing them lifelong lessons. This in-home Makerspace has many benefits over the previous two, the primary of which being that you are the boss. The long list of bureaucratic hoops you need to jump through to establish your space disintegrate when you happen to own it. Your only restrictions are the legality and safety of the space itself.

Make sure you understand the legal implications of having individuals outside of your family work your Makerspace, regardless of where it is located. Review your local and state laws regarding your liability with respect to those working in your Makerspace so that everyone is covered in the event of an accident.

The idea of outfitting a garage to function as a Makerspace can get a bit tricky. We are often faced with poor insulation and unfinished walls, missing HVAC vents, sparse electrical connections and last but not least, cars. Yet, with all of the potential drawbacks, the idea of converting a garage into a space for creation and making it available to the desired community is extraordinary.

Project: Equipment Donations and Discounts

If your Makerspace is to function as a nonprofit organization, there is a good likelihood that you can solicit equipment donations or a discount to the listed price from outside organizations. Donations can also qualify as tax-exempt if your Makerspace qualifies under section 501(c)(3) of the United States Internal Revenue Code or the donation is made to the qualifying library or school in which you are established. This project is designed to assist you in determining what equipment, materials, and so on you think could be donated as well as provides a strategy for contacting the potential donor.

You can find out more about how your Makerspace might qualify as tax-exempt by visiting the IRS's website at www.irs.gov.

Regardless of your tax-exempt status, the nature of the Makerspace is attractive to outside organizations because it serves as a hub for inspiring future engineers. Many organizations would love to either sponsor or donate items to your Makerspace if they see that their resources will be used for the betterment of the community. This project is designed to walk you through the preliminary steps for soliciting a donation or discount to items for your Makerspace and helping to alleviate potential financial burden.

Materials

Provided in the following sidebar is an "Industry Contact Form" that walks you through all the steps necessary to research a specific needed component, contact a potential supplier, and present your project in a way that illustrates to the organization the benefits of the donation or discount. Good luck!

Procedure

Step 1

Start by researching the item that you believe could be obtained through donation. The objective is to conceptualize equipment or materials whose donation would ultimately benefit the company. The potential donor might have overstock inventory, damaged goods, or a stockpile of materials headed to the dump, all of which are gold for a Makerspace. These items end up consuming space and therefore resources, making them prime for donation. Be sure to record the company name and contact information on the Contact Form during your research.

Step 2

After you have determined your item, its important to record specifications that can help the potential donor understand what you are requesting. Make sure you focus on specifications like model, part number, dimensions, quantities, color, and any other specification that you require. This helps the company spend less time trying to understand your request and will aid them in directing you to the individual who can better assist you. If you do your homework, the results will be evident.

Step 3

Time to make the call. Making a phone call rather than sending an email is not only more professional, but shows that you are taking the time to make the request in person. It is a lot harder to ignore a phone call than it is to delete an email. The Contact Form contains example dialogue that will assist you in your request. Good luck with you phone call and your potential donation/discount!

Designing the Space

Whether its nestled in a home basement or spread across a warehouse, the design and layout of your Makerspace is the most important initial decision you can make. It will ultimately dictate the ability of its Makers to work efficiently and safely as well as shedding light on the types of equipment it can sustain. The types of work surfaces, shelving, computer desks, and seating all play a part in the flow of your space and are elements that can be visualized prior to actually being installed. All three locations previously discussed have predetermined layouts with little room for modification. Unless you have the authority to put a new window in the side of a library or a classroom, or building permits to poke a hole in the side of your house, we end up working with what we have. It is now up to your own ingenuity to shape and mold your future Makerspace into a form that not only supports your equipment, but provides an effective work environment for its occupants.

Check with your local government authorities to verify the need for building permits when modifying your existing structure. Tasks as small as moving an outlet can require a permit.

Makerspace Industry Contact Form

Illustrating the mission of your Makerspace to the layman is no easy task, especially if they are unfamiliar with the Maker movement and its associated technologies. If your Makerspace requires new equipment or materials, you might be able to solicit a donation or discount by directly contacting industry representatives. Sources of assistance can range from local businesses to online distributors; the possibilities are truly endless. By making a phone call in lieu of writing an email, you are establishing a more personal relationship that is harder to disregard, and who knows, you might end up with a continuing sponsor. During this conversation, you will be briefly describing your project and the component you require in a manner that is professional and reflects the legitimacy of your project.

Research the Donation

Prior to making your phone call, it is important to know the exact specifications of the equipment or materials you are requesting (Table 1-4). This helps to limit the amount of explanation required when making your contact. The rule of thumb to follow is: time is money. If you lack confidence with your request and ramble or are unclear, the chances your request is successful dramatically decrease.

Table 1-4. Company and equipment description

Company Name	Phone Number	Website	Component Description
Part Name	**Part Number**	**Size/Qty**	

Make the call

Introduce yourself professionally; for example, "Hello, my name is _____ , and I am establishing a Makerspace workshop for my local community. This workshop is designed to educate today's youth through the use of open source software and technology."

Briefly describe your intentions; for example, "For our Makerspace to succeed and provide the most effective environment for our patrons, we need a _____. Would your company be willing to sponsor us by donating or offering a discounted price to this equipment/material?"

They will either say "No," and you say "Thank you for your time," and you should try another company.

Or, they say something along the lines of, "Yes, let me transfer you to the _____ dept." and you start the process again.

Make sure you use *Please, thank you, yes Ma'am/Sir*. You might think this is not necessary, but you would be surprised at how much being polite can help.

Complete the following during your phone call (Table 1-5):

Table 1-5. Conversation log

Contact Name	Extension	Successful Donation Y/N
Details		

Remember, if you receive a donation or discount, make sure you provide a written "thank you" letter to the donor. This not only maintains good rapport with your potential vendors, but aids in establishing a continuing relationship.

This chapter discusses a variety of design considerations and rules that can be followed to ensure the success of your new Makerspace. Although there are a number of ways to begin designing your space, the objective is always the same: pPlan first, and then act.

Nobody likes to "rub elbows" while they are working on their project. It is especially important when working with machines or potentially dangerous tools that enough free space is available at all times. This safety zone ensures that multiple people can be working in the same general area while maintaining a safe operating distance. A good rule of thumb for establishing safety zones is to give 3 sq ft of space around anyone using a hand tool, tabletop drill press, or soldering iron, and 3 ft radius around any standing piece of equipment (for example, a band saw, standing drill press, or lathe). Tools like a horizontal band saw or table saw will require more space to accommodate for the use of large materials.

Work Areas

You can never have enough space, especially when it comes to work areas. These spaces should cater to both the project and the personality type of the Maker. Some people like to think aloud and spread their projects to the limits of the work area, whereas others tend to work with more methodology and organization.

> *Connect power tools and equipment to outlet timers. This helps eliminate problems related to equipment being left on after use. Just make sure they are rated at a high enough power level.*

One method for catering to these differing work methods is to designate quiet and loud work areas. The quiet work area should facilitate project types that don't involve a great deal of mess or noise. This type of work area is actually better at accommodating more people than the loud area because the environment is less physically hazardous. The loud work area should be able to handle larger, more intensive project types. These could require power tools and pneumatics, thus requiring less people per square foot so as to reduce the potential for injury.

Each work area should facilitate a specific type of project and should contain all of the resources necessary to operate in that area. Makerspaces commonly have the areas listed in Table 1-6, and vary depending on the direction of the Makerspace.

Protect your work surfaces (see Figure 1-8). Regardless of your Makerspace's location, it is important to keep your surfaces clean. An easy way to accomplish this is by covering the surfaces with Masonite. This material, which is what clipboards are typically made of, provides a smooth and resilient surface that is relatively heat and moisture resistant, not to mention inexpensive.

Figure 1-8. *Covering work surfaces with inexpensive and durable materials helps to prolong the life and quality of the work surface.*

Table 1-6. Makerspace work areas

Work Area	Space Requirement	Power Requirement	Primary Function
Computer Station	Small	Large	Supports computers, printers, and common software

Work Area	Space Requirement	Power Requirement	Primary Function
Soldering Station	Small	Small	Contains necessary soldering equipment, ESD protection, and support electronics
Electronics Workbench	Medium	Small	Houses electronics components and test equipment
Project Area	Large	Large	Open floor and project work surfaces
3D Printing Station	Small	Small	Supports 3D printer, computer, and filament
Laser Engraving Station	Medium	Medium	Supports laser engraver, ventilation and laser-able materials
Power Tool Area	Large	Large	Supports standing and portable power tools, open floor, and work surfaces

Figure 1-9. *The computer station not only contains a computer, but has a small amount of open work space for writing and working on projects.*

Devices like electrical ceiling drops (mainly for the garage and school Makerspace) and extension cord covers should be used in lieu of exposed cords as they greatly improve work flow and personal safety.

Every Makerspace needs computers. They act as the gateway for project research, design, and control as well as provide support for many pieces of equipment. The main decision behind establishing designated computer stations (Figure 1-9) is whether you will be providing desktop or portable computers. Each has their advantages yet require different infrastructure to support.

If you choose to use desktop computers, which are often available in a library or school environment, it is necessary to allocate enough surface area to support the computer and its peripherals. Depending on the size of the computer systems, a 6 ft × 2 ft table is capable of supporting two computer stations comfortably. The benefit of having dedicated computer stations is a more consistent work environment. You know where your computers are at all times, and can easily account for the necessary space required for their use.

Following the portable route, you will need to take into account the need for charging the systems or provide ample access to electrical outlets at the different workstations. The advantage of using portable computers in this environment is the lack of need for dedicated computer work surfaces. This area can now be repurposed to support more equipment, soldering stations, and so on. However, there is a pretty big drawback with having a large quantity of power cables traveling around the space, which could result in a tripping hazard.

Project: Makerspace Layout Tool

The initial inclination when designing a space is to sit behind a computer and lay it out by using AutoCAD. Although this is an effective means for designing a space, physically laying it out by hand can yield better results (see Figure 1-10). There is something to be said about physically connecting with your design and the flexibility of using nonvirtual materials to visualize the environment. This project is designed to help you begin the layout process by analyzing your available space, tools, and materials in a way that produces an effective design for your Makerspace.

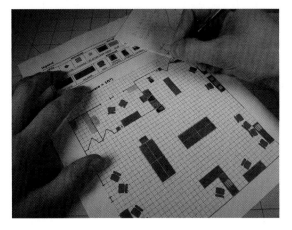

Figure 1-10. *Print out the layout tool and cut out the examples to begin designing your space.*

Materials

You can find the files for this project on Thingiverse.com (*http://bit.ly/1asvrtV*).

Equipment Checklist		
Type	**Size (ft)**	**Power (Watts)**

Procedure

Step 1

To begin, you need to determine how much space you actually have available. Starting at one corner of the room, begin to measure the wall length and illustrate the wall on the Layout Paper. Remember, each cell represents a 1 ft × 1 ft area. Indicate the location and width of any doors, windows, or other obstruction that influence the spaces design and work-surface placement. Continue this process until the perimeter has been fully outlined.

Step 2

Reference two common walls and use your tape measure to determine an approximate X/Y coordinate for any feature located internal to your space and illustrate them on the Layout Paper. Be sure to include electrical outlets, support features (load-bearing poles, interior walls), floor drainage, and lighting. These features are important to note because they will set the constraints for equipment placement.

Step 3

Record the tools, materials, and workspaces you currently have or would like to have in your Makerspace. Using your tape measure, record the footprint dimensions of your current equipment and illustrate them on the supplied Available Equipment Checklist. It is also a good idea to consider growth and research the dimensions of equipment you would like to acquire at a future date. This will help you better organize the initial layout of your space (Figure 1-12) and will make future expansion easier.

Step 4

You should now have an accurate representation of your space's structural components and available utilities and can begin to determine how your equipment, materials, workspaces, and storage might be organized. Using the Available Equipment Checklist and Sample Equipment Sheet, cut out

legend

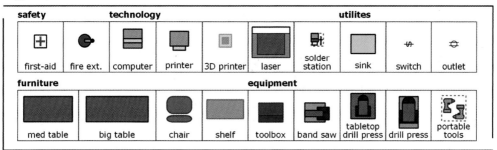

grid space (1 square = 1 ft²)

Figure 1-11. *Makerspace Layout Tool.*

and begin to position them in your space. Make sure you account for work-zone safety rules, position equipment near their required utilities, and envision how the members might utilize the space.

Safety

Not only should safety be the main concern when designing and laying out your Makerspace, but it should be a topic that is often addressed during its operation. Whenever students enter into a laboratory environment and are expected to use equipment, they are required to pass a series of Safety Tests that are designed to verify competency in the safe operating procedures of both the space and its equipment. With respect to the more public environment of the library and garage Makerspace, it is imperative that an understanding of the safety rules and regulations your Makerspace follows are clearly understood.

The more you encourage safe practices, the less likely you will have to deal with damaged equipment and tools, poor work conditions, and, in the worst case scenario, injury. So, take the time to il-lustrate how you want your Makerspace to operate and ensure that everyone conforms to your rules.

Safety Tests

A great way to welcome new Makers into your Makerspace is with a Safety Test! Many of us cringe at the sight of written assessment, but rest assured, these tests are a little different. It is often difficult to assess skill sets, especially when it comes to the safe use of equipment. Regardless of what someone says about their previous exposure and use of any of your Makerspace's equipment, it is mandatory that you administer some form of safety assessment and the test taker receive a score of 100 percent. These safety tests are used to both evaluate the member's written knowledge regarding the proper use of equipment as well as her ability to physically demonstrate her skill. The goal of these tests are not to prohibit the equipment's use; rather, they are to shed light on who needs guidance until they reach a skill level at which they can operate autonomously. The following safety tests can be used to help maintain a safe working environment for you and your space's members.

General Makerspace Safety Test

The following test is designed to assess your comprehension of the safe operating practices required by our Makerspace. Answer the questions to the best of your ability.

True/False and Explain
1. T/F: Safety glasses and safety goggles offer the same type of protection.

    a. Answer:

    b. Explain:

2. T/F: Protective eye covering should only be worn when working with tools that are potentially dangerous.

    a. Answer:

    b. Explain:

3. T/F: Small cuts and scrapes do not require the attention of the instructor/supervisor.

   a. Answer:

   b. Explain:

4. T/F: It is acceptable to leave a work surface cluttered while not in use.

   a. Answer:

   b. Explain:

5. T/F: Personal items such as book bags can be left on the floor while you work.

   a. Answer:

   b. Explain:

6. T/F: Tools should only be used for their intended purpose.

   a. Answer:

   b. Explain:

7. T/F: Electronic devices such as soldering irons should be left plugged in if someone else is waiting to use it.

   a. Answer:

   b. Explain:

8. T/F: Tools should be kept in an organized location.

   a. Answer:

   b. Explain:

9. T/F: Help should be requested whenever large materials, equipment, or work surfaces need to be moved.

   a. Answer:

   b. Explain:

10. T/F: When stocking a shelf, the heaviest items should be placed on the bottom.

   a. Answer:

   b. Explain:

Completed by:

Date:

Checked by:

Date:

General Makerspace Safety Test Answer Key

1. Safety glasses and safety goggles offer the same type of protection.

   a. **Answer**: True

   b. **Explanation**: Both are designed to protect your eyes from foreign objects.

Glasses tend to work better for individuals who do not currently wear glasses, wheras goggles tend to be more comfortable. The important feature to note is the glasses impact rating. Look for a "+" mark on the lens, which indicates the glasses conform to high-velocity standards.

2. Protective eye covering should only be worn when working with tools that are potentially dangerous.

   a. **Answer**: False

   b. **Explanation**: Just because you are not working with a potentially dangerous tool, doesn't mean that you cannot be injured. If anyone in the room is working with a tool that could potentially be dangerous, everyone must wear eye protection.

3. Small cuts and scrapes do not require the attention of the instructor/supervisor.

   a. **Answer**: False

   b. **Explanation**: Regardless of the severity of the injury, the instructor/supervisor must be notified. This is important because it helps to maintain a clean and controlled work environment.

4. It is acceptable to leave a work surface cluttered while not in use.

   a. **Answer**: False

   b. **Explanation**: Work surfaces should be kept as clean as possible so as to avoid potential injury.

5. Personal items such as book bags can be left on the floor while you work.

   a. **Answer**: False

   b. **Explanation**: Personal items, especially those with straps, pose a serious tripping hazard.

6. Tools should only be used for their intended purpose.

   a. **Answer**: True

   b. **Explanation**: By using a tool as it was intended, the lifespan of the tool is dramatically increased. Common mistakes include using screwdrivers as pry-bars, wrenches as hammers, and hex keys on Torx bolts.

7. Electronic devices such as soldering irons should be left plugged in if someone else is waiting to use it.

   a. **Answer**: False

   b. **Explanation**: Safety should never be sacrificed in an effort to save time.

8. Tools should be kept in an organized location.

   a. **Answer**: True

   b. **Explanation**: This helps to maintain an orderly work environment.

9. Help should be requested whenever large materials, equipment, or work surfaces need to be moved.

   a. **Answer**: True

   b. **Explanation**: Following this practice ensures the safe relocation of said objects. Trying to move them by yourself could result in the unintentional injury to yourself or those around you.

10. When stocking a shelf, the heaviest items should be placed on the bottom.

    a. **Answer**: True

    b. **Explanation**: Following this rule helps to prevent unintended injury due to items falling from an elevated surface.

Hazardous Materials

Virtually every material and compound has been analyzed and a material safety data sheet (MSDS) has been generated by its manufacturer to ensure safe handling. Even water has an MSDS! Each sheet illustrates the material's chemical identification, hazard identification, composition, emergency and first aid measures, fire-fighting measure, accidental release measure, precautions for safe handling and storage, exposure control measures, physical and chemical properties, stability and reactivity data, and the list goes on. So how can we use an MSDS to better understand how to ventilate the Makerspace? Well, it's quite easy.

ABS, the common feedstock for 3D printing, has an MSDS that reports all of the potential hazards encountered when handling the material. During printing the feedstock is melted and in turn releases volatile organic compounds (VOCs) into the air, which need to be contained in order to provide a safe and healthy work environment. The sheet for ABS states the following:

Hazards Identification

Vapors and fumes from heat processing may cause irritation of the nose and throat, and in cases of overexposure can cause headaches and nausea. If affected, remove to fresh air and refer to a physician for treatment.

Exposure Controls/Personal Protection

Local exhaust at processing equipment to assure that particulate levels are kept at recommended levels.

So you see from the MSDS for ABS that it is necessary to appropriately ventilate the exhaust fumes from your 3D printer (as shown in Figure 1-12) in order to maintain a healthy and safe environment. Nobody wants to barf on their printer, right?

Figure 1-12. *Soldering fume extractors can be used for ventilating the 3D printer when printing materials like ABS.*

Some of the materials you might encounter in your Makerspace, primarily solder paste, require refrigeration. This can pose a huge health risk if these materials are stored in the same refrigerator where food is kept. It is *highly* advised that you acquire a separate refrigerator and clearly label it as "NOT FOR FOOD ITEMS!" and add a Mr. Yuck poison control sticker.

Other materials might need to be stored in an appropriate flame-resistant cabinet (Figure 1-13). This cabinet is designed to isolate both the inside of the cabinet from a fire in the room, should one occur, and the outside of the room if something inside the cabinet were to ignite.

Figure 1-13. *Flammables should be stored in an appropriate cabinet to help prevent unintended fire.*

The general rule when working with potential hazardous materials is to fully understand the extent of the hazard before handling. Every Makerspace should have a hardcopy set of MSDS sheets for every hazardous material present in the space and the phone number for Poison Control readily available. When it comes to safety, you can never go too far. The integrity of your space and the well being of its members is crucial to its success.

Ventilating Your Equipment

Many of the tools found in a Makerspace produce potentially harmful fumes while in operation. Soldering, 3D printing, and laser cutting are the main sources for unwanted fumes, and it is important to employ a proper method for ventilating the area. Generally, it is necessary to have at least one source of fresh air for your Makerspace, whether it is a window, roof access, or pre-existing ventilation system. Always refer to your state and local building codes when choosing and installing any ventilation system.

Ventilating Your Soldering Station

When you begin to solder, you will notice a fair amount of white smoke rising from your project. *Don't worry!* This is normal and doesn't mean that you have released the dreaded "white magic smoke" that occurs when a component is overpowered or short-circuited. The smoke that you

see is released from the flux that is present in the solder (this is discussed in detail in Chapter 2.

> *Each manufacturer uses a different flux formulation for their solder. It is important to acquire the appropriate MSDS sheet and be aware of the risks involved with handling and exposure to the flux vapors.*

Even though the short-term exposure to solder fumes is not usually an issue, it is better to be safe than sorry. This is especially important if the Makerspace is in a room without proper ventilation or if there is a concern for anyone with a respiratory problem. One of the easiest ways to tackle flux fumes is to use a localized fume extractor like the Weller WSA350. They are portable, relatively inexpensive, and are great at tackling fumes. An extractor such as this relies on an activated carbon filter and a high-powered fan to extract noxious components from the air.

Ventilating Your 3D Printer

3D printers can produce fumes that cause headaches and nausea if overexposed and thus require fume extraction so as to provide a safe working environment. This is not necessary for most high-end printers such as those from Stratasys because they handle fume extraction internally. Yet for the vast amount of low-cost 3D printer designs and sizes, it is necessary to install a system with some degree of flexibility.

This system can be as simple as locally installing a fume extractor, like the tabletop model as mentioned in the previous section, or positioning the 3D printer near a window or other access point. In the case of having your printer near a window, it is important to ensure that the Makerspace has positive air-flow so that the fumes don't come back into the room.

A good way to test the proper functioning of your fume extractor is to touch a small amount of flux cored solder to the tip of a soldering iron near the location of your fume source while the extractor is turned on. You can then watch the path of the smoke as it rises and adjust the position of the extractor to intercept the vapors.

Ventilating Your Laser Cutter

The most overlooked part of purchasing a laser cutter is its ventilation system. This system is designed to provide a vacuum effect on the material bed as well as flushing the machine of material smoke and fumes. Each laser cutter has specific cubic-foot-per-minute (CFM) ventilation requirements that can result in a pretty sizable system. Generally, the system consists of a series of ductwork that extends from the back of the laser cutter and connects to a blower fan. This fan is then connected to duct work that is fed into an internal duct system or directly outside.

Project: Directed Smoke Absorber

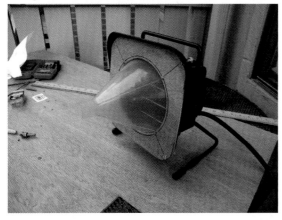

If you find that your fume extractor cannot be properly positioned, it can be modified to localize the extraction point closer to your work. With a few simple tools and materials, this project will walk you through the steps to create a hose attachment for your WSA350. Although this project can be completed by hand, having access to a laser cutter makes is significantly easier!

This project requires safety glasses, which should be worn for the entirety of the project.

Materials

Materials List		
Item	**Quantity**	**Source**
Soldering smoke absorber	1	electronics distributor
6 in diameter plastic funnel	1	automotive parts store
1/8 in plywood sheet	1	home improvement store
6-32 × 2 in bolt	3	home improvement store
6-32 nut	3	home improvement store
#6 washer	3	home improvement store
2 in washing machine or sump pump utility hose	1	home improvement store
2.5 in hose clamp	1	home improvement store
1/2 in adhesive-backed insulation foam	1	home improvement store

Makerspace Tools and Equipment
1/4 in drill bit
1/8 in drill bit
C-clamp
Center-punch
Coping saw or equivalent
Power drill

Procedure
Step 1

Start by tracing the outline of the absorber onto the plywood sheet. Mark four holes approximately 1 inch in from the four corners. These will be used for securing the plywood to the absorber. Find the center of the of the outline and trace the shape of the funnel. Create a second circle concentric with the funnel outline that is approximately 1/4 inches smaller. This will allow for the funnel to be mounted to the plywood. Mark three holes around the perimeter of the outer funnel outline. These will be used for securing the funnel.

Step 2

Clamp the plywood sheet to the table by using the C-clamp and a scrap piece of wood. Then, using the center-punch, mark the center of the three holes by pressing the center-punch slightly into the material. Attach the 1/8 in drill bit to your drill and drill holes for the funnel mount. Make sure that you are drilling into an appropriate surface! Remove the bit, secure the 1/4 in drill bit to your drill, and drill out the mounting holes for the two brackets.

Step 3

Using the coping saw (Figure 1-14), carefully cut around the outlines you previously marked. Test fit the plywood plate onto your smoke absorber and trim away any excess material. Don't worry if you cut away too much material because you are going to seal the back of the plate with insulation foam.

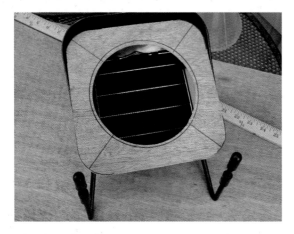

Figure 1-14. *Use a coping saw or jig saw to cut out the large hole in the center and outline of the plywood plate.*

Step 4

Mark a 1/4 in offset line around the back side of your plate. This line will act as a guideline for the insulation foam. Attach the insulation foam (Figure 1-15) to the back of the plate so that it follows the guideline. Secure the funnel onto the plywood by using the three 6-32 fasteners. Place the assembly onto the absorber and secure in place by using the 1/4 in fasteners.

Figure 1-15. *Attach foam tape to the back of the plywood plate to prevent any air leaks.*

Step 5

Attach the hose to the hose adapter and fire up your smoke absorber. If everything is installed correctly, the vacuum produced should hold the plate in place and there should be a good vacuum now at the end of the hose. Position this hose as close to the fume source as necessary and secure in place with the hose clamp. Congratulations! You should now have a much more effective localized smoke absorber!

Tools | 2

IN THIS CHAPTER

- Tool Safety
- Hand Tools
- Power Tools

A Makerspace can never have enough tools. There will always be a new ratchet, precision screwdriver, or lithium iron phosphate-powered drill that can be added to the collection. Some tools are designed to operate around the forces produced by the human body, whereas others require external power (Figure 2-1). Some separate material, others join them together. Tools make it possible for us to create, destroy, hack, and manipulate the world around us. Having the right tool for the job make these processes infinitely easier.

Figure 2-1. *A power tool in use.*

Developing a good understanding of the proper use and care of the tools in your Makerspace is important for maintaining a functional work en-

vironment. Sure, you could use a screwdriver as a chisel or drill a wood bit through concrete, but the result will be subpar, and you might break the tool in the process. So how do you know what tool is right for the job? Well, the answer is easy. Each tool is designed to perform a specific task, whether it grips, strips, bashes, or pries and will do so for a long time. When tools are misused, the chance of damage to the tool or the workpiece is greatly increased and poor results will follow. The list that follows should serve as a quick guide for choosing the appropriate tool for the job.

Which tool do I use?

- Removing a big bolt from the wheel of a robot

  — **DO** use a socket that fits the head of the bolt and a socket wrench to remove it. Alternatively, an adjustable wrench can be used, as long as not too much torque is applied, which will result in damage to the head.

  — **DO NOT** use a pair of pliers to remove the bolt. Pliers are only meant to hold work in place, not wrench it. The sharp teeth in the pliers jaws will cause damage to the head of the bolt when turned.

- Removing a nail from a board:

— **DO** use a claw hammer or pry bar to remove the nail by seating the head of the nail into the hammer's claw and carefully prying the nail out.

— **DO NOT** use a screwdriver, pliers, scissors or teeth to try to remove a nail. Enough said.

- Removing a stuck gear from a motor shaft:

  — **DO** use a gear puller to remove a gear from a shaft. Many gears are press fit onto shafts and require slow, even pressure to remove. Alternatively, the gear can be heated until it expands, allowing for easy removal but care must be taken not to damage areas around the heated gear.

  — **DO NOT** use a hammer or pry bar to remove the gear. The shaft of the motor is connected to two bearings and a magnet/coil depending on motor type. When a large force is exerted onto the shaft it *will* break the components inside of the motor resulting in permanent damage.

- Drilling a 1 in hole into thin metal:

  — **DO** use a center-punch to mark the hole followed by 1/4 in pilot hole. Next gradually increase the bit size until the final diameter is reached. The pilot hole reduces the amount of material a 1 in bit would remove, making the process both easier and safer. If the material is too thin to drill (<3/16 in thick) a hole punch should be used.

  — **DO NOT** use a 1 in bit to drill a 1 in hole. Without the removal of material prior to drilling the hole, the bit will bind causing the drill to stall, greatly increasing the potential for injury.

- Securing a crimp connector to wire:

  — **DO** use a wire stripper to remove approximately 1/4 in of insulation from the end of the wire. Place the crimp connector over the wire and crimp in place by using the designated notch on the crimper's jaw.

  — **DO NOT** strip more then 1/4 in of insulation and leave excess wire exposed. This uninsulated wire can cause electrical shorts and can potentially damage the equipment or cause personal injury.

Tool Safety

Tools are designed to amplify and extend the abilities of the human body. In doing so, the potential for personal injury to you and those around you is greatly increased. Developing proper understanding of a tool's use and its abilities will help reduce the potential of an accident, and if one occurs, it can help limit its impact.

Everyone working in the Makerspace should be trained on each piece of equipment and should understand not only the proper use of the tool, but its potential hazards. Following these simple precautions will help to maintain a safe working environment.

General Tool Safety Rules

1. Always wear safety glasses, even if you are not working directly with the tool.

2. Never carry a tool by the cord.

3. Use a clamp to secure your work rather than your hand.

4. Wear closed-toe shoes if working with portable power tools or other unwieldy objects.

5. Only use a tool for its intended purpose. If you are unsure of what the proper tool is, ask someone who is.

6. Keep your work area clean and free of tripping hazards.

Hand Tools Safety Test

The following test is designed to assess your comprehension of the safe operating practices required by our Makerspace. Answer the questions to the best of your ability.

True/False and Explain

1. T/F: It is alright to use your hands to hold your work in place when lightly filing or sanding.

    a. Answer:

    b. Explain:

2. T/F: You can use a slotted screwdriver as a pry bar.

    a. Answer:

    b. Explain:

3. T/F: When handing someone a sharp or pointed tool, you should do so handle-first.

    a. Answer:

    b. Explain:

4. T/F: Sharp tools can be less dangerous than dull tools.

    a. Answer:

    b. Explain:

5. T/F: The appropriate solution for removing a stuck fastener, such as a nut and bolt, is to hit it with a hammer.

    a. Answer:

    b. Explain:

6. T/F: It is alright to cut with a knife toward yourself, as long as you are careful.

    a. Answer:

    b. Explain:

7. T/F: You can use standard sockets on metric bolts, and vice versa.

    a. Answer:

    b. Explain:

8. T/F: You can use hydraulic jacks as permanent support.

    a. Answer:

    b. Explain:

9. T/F: It is alright to throw razor blades and other sharp objects in the trash.

    a. Answer:

b. Explain:

10. T/F: You can use scissors to cut virtually any material.

   a. Answer:

   b. Explain:

Completed by:

Date:

Checked by:

Date:

Power Tools Safety Test

The following test is designed to assess your comprehension of the safe operating practices required by our Makerspace. Answer the questions to the best of your ability.

True/False and Explain

1. T/F: Long hair, drawstrings, and neckties should be secured prior to using a power tool.

   a. Answer:

   b. Explain:

2. T/F: You are required to request the permission of the instructor/supervisor prior to using a power tool.

   a. Answer:

   b. Explain:

3. T/F: It is alright to use one hand to hold your work and the other to operate the tool.

   a. Answer:

   b. Explain:

4. T/F: A vise and a C-clamp are two tools that you can use to secure your work in place.

   a. Answer:

   b. Explain:

5. T/F: Tool guards and shields are designed to protect you from being cut.

   a. Answer:

   b. Explain:

6. T/F: It is OK for two people to work with the same tool.

a. Answer:

b. Explain:

7. T/F: Chuck keys should be left on top of the machine when in use.

a. Answer:

b. Explain:

8. T/F: It is important to notify those around you prior to the use of a loud power tool.

a. Answer:

b. Explain:

9. T/F: The fence or blade guide can be adjusted while the tool is in operation.

a. Answer:

b. Explain:

10. T/F: It is alright to set the tool down while the motor is still running, as long as there is a guard over the blade.

a. Answer:

b. Explain:

Completed by:

Date:

Checked by:

Date:

Safety Test Answer Keys

Here are the answer keys for the preceding tests.

Hand Tools Safety Test Answer Key

1. It is alright to use your hands to hold your work in place when lightly filing or sanding.

    a. **Answer**: True

    b. **Explanation**: This is generally safe to practice. If the sanding or filing is of a dense material, it should secure the material appropriately.

2. You can use a slotted screwdriver as a prybar.

    a. **Answer**: False

    b. **Explanation**: Prybars should be used as prybars. Screwdrivers are meant to drive screws.

3. When handing someone a sharp or pointed tool, you should do so handle-first.

    a. **Answer**: True

    b. **Explanation**: This helps to ensure that the person to whom you are handing the tool does not get injured by the sharp or pointed tip.

4. Sharp tools can be less dangerous than dull tools.

    a. **Answer**: True

    b. **Explanation**: Everything is relative, but in general, a sharper tool requires less energy to cut and results in a more controlled use. Dull tools are often over worked when cutting, and you have a higher likelihood of losing control.

5. The appropriate solution for removing a stuck fastener, such as a nut and bolt, is to hit it with a hammer.

   a. **Answer**: False

   b. **Explanation**: Although it would work, stuck fasteners should be removed by using the appropriate means. Using a hammer can result in damage to the surrounding area or unintended injury.

6. It is alright to cut with a knife toward yourself, as long as you are careful.

   a. **Answer**: False

   b. **Explanation**: Regardless of the scope of the cut, cutting tools should always be used so that the cutting motion is directed away from you and those around you.

7. You can use standard sockets on metric bolts, and vice versa.

   a. **Answer**: False

   b. **Explanation**: Using the incorrect socket can result in damage to the nut, bolt head, or the tool itself.

8. You can use hydraulic jacks as permanent support.

   a. **Answer**: False

   b. **Explanation**: Hydraulic jacks are only to be used to elevate an object. Permanent support should be by means of a jack stand or other load-bearing stable support.

9. It is alright to throw razor blades and other sharp objects in the trash.

   a. **Answer**: False

   b. **Explanation**: Placing sharp objects in the trash can result in injury to the person handling disposal. These objects should either have their edges concealed in an appropriate container or other means prior to disposal.

10. You can use scissors to cut virtually any material.

   a. **Answer**: False

   b. **Explanation**: Everything is relative. It is important to use a cutting tool as it was intended. Scissors for light materials, shears and snips for heaver materials, and so on.

Power Tools Safety Test Answer Key

1. Long hair, drawstrings, and neckties should be secured prior to using a power tool.

   a. **Answer**: True

   b. **Explanation**: This rule is of utmost importance to follow because these unsecured items can be quickly entangled in power tools.

2. You are required to request the permission of the instructor/supervisor prior to using a power tool.

   a. **Answer**: True

   b. **Explanation**: This rule helps to maintain a safe working environment by making the supervisor aware of the tools being used and the qualifications of those using them.

3. It is alright to use one hand to hold your work and the other to operate the tool.

   a. **Answer**: False

   b. **Explanation**: Regardless of the power tool, your work should be mechanically secured to a surface to prevent injury and reduce the potential of your work becoming a projectile.

4. A vise and a C-clamp are two tools you can use to secure your work in place.

   a. **Answer**: True

   b. **Explanation**: Even when using these tools, it is important to ensure that the work surface is secure, and in the case of the vise, it should be securely fastened to the work surface.

5. Tool guards and shields are designed to protect you from being cut.

a. **Answer**: True

b. **Explanation**: These devices are designed to prevent you from being cut. These should not be removed in an attempt to create more work area or speed up tool use.

6. It is OK for two people to work with the same tool.

   a. **Answer**: False

   b. **Explanation**: It is impossible to gauge the reaction of someone else, especially if they are using the same tool. There are very few cases in which this is allowed; for example, when you are assisting with large material on a table saw.

7. Chuck keys should be left on top of the machine when in use.

   a. **Answer**: False

   b. **Explanation**: Chuck keys should be securely stowed while not in use. Leaving a chuck key on top of a machine can result in it becoming a projectile or distracting the machines operator if it falls off.

8. It is important to notify those around you prior to the use of a loud power tool.

   a. **Answer**: True

   b. **Explanation**: By informing those around you that you are about to make a loud noise helps reduce the risk of them being startled and potentially injured as a result.

9. The fence or blade guide can be adjusted while the tool is in operation.

   a. **Answer**: False

   b. **Explanation**: These safety features should only be adjusted when the machine has come to a full stop. Unintended interaction with the blade might occur otherwise, causing damage to the machine and potential injury to its operator.

10. It is alright to set the tool down while the motor is still running, as long as there is guard over the blade.

a. **Answer**: False

b. **Explanation**: Blade guards are only a safety feature and can still fail. Serious injury can occur if the rotating blade comes in contact with the work surface, or if someone touches the blade while it is still rotating.

Hand Tools

Hand tools are the bread and butter of any Makerspace. They don't require external power, are relatively safe to use, and can be quite inexpensive. Armed with the right array of hand tools, your Makerspace will have the ability to tackle virtually any construction or deconstruction task. This section covers tools that are designed to separate, join, bend, and manipulate materials, ranging from balsa wood to carbon steel.

Look for tools that feature a lifetime warranty. This warranty often covers misuse and more than pays for the cost of the tool when redeemed.

Screwdrivers

The screwdriver is a tool that is used to apply a turning force to a screw with a matching drive system. These drive systems are designed to allow enough torque to secure or remove a fastener while limiting damage to the mating interface. For example, systems such as Phillips are designed to "cam-out" the driver when too much torque is applied to the screw. This in turn prevents damage, or rounding of the mating surfaces.

The common screwdrive features are identified as follows:

Handle
   Originally constructed out of wood, the screwdriver's handle is designed to create a torque-bearing interface between the human hand and the shaft of the screwdriver.

Today, screwdriver handles are made primarily out of cellulose acetate, a clear polymer manufactured from wood-derived cellulose. Handles can also feature exotic mechanisms and ergonomic designs that afford better control and usability.

Shaft

A screwdriver's shaft is designed to provide a mechanical interface between the handle and the tip. Shafts are made out of steel and plated with nickel, chromium, or titanium to provide protection from corrosion and add strength.

Tip or Bit

The tip is designed to transfer the input torque to the driven fastener. Tip manufacturers often employ features that improve upon standard tip performance. These methods include heat treating for strength, magnetizing the tip for convenience and adding ribs that resist camming-out under heavy torque.

Screw Drive Systems

Each drive system has a sizing convention that is used to identify the size of the drive and the mating fastener. Table 2-1 illustrates the common sizes of each drive system.

Table 2-1. Common screw drive types and sizes

System Name	Tip Number	Screw Size
Hex	0.05 in–1/2 in and 0.71 mm–12 mm	Same as Tip Number
Phillips	#000–#4 and 1/8 in–9/32 in	#000–#12 and 1/4 in–1/2 in
Slotted	#0–#3 and 7/64 in–3/8 in and 0.8 mm–4mm	#000–#16 and 1/4 in–1/2 in
Square	#00–#4	#1–#14 and 1/4 in–1/2 in
TORX(r)	T1–T55	Same as Tip Number

Hexagonal

Similar to the hexalobular drive, the hex drive system uses a hexagonal drive interface (Figure 2-2). This design reduces cam-out problems and yields a better torque transfer over Phillips or slotted systems. Due to the shallow interface angle between the driver and the screw, the hex drive system has a tendency to strip or round off the contact points under excessive torque loads.

Figure 2-2. Hexagonal drive.

Phillips

The Phillips screw drive system is designed to self-align the driver into the head of the fastener and provides continued contact even when misaligned (Figure 2-3). When too much torque is applied to the head of the fastener, the system is designed to cam-out the driver, preventing permanent damage. One of the most common mistakes when using Phillips screws is to use an incorrectly sized drive. This can cause damage to the fastener and if severe enough, will require specialized tools to remove it.

Figure 2-3. Phillips drive.

Slotted

The slotted screwdriver is the earliest form of a screw drive system (Figure 2-4). This tool is made by drop-forging and grinding the tip of the shaft into a chisel shape.

Figure 2-4. Slotted drive.

Square Drive

The square drive system is designed to not cam-out under high torque loads (Figure 2-5). Because of this feature there is very little wear at the drive interface, making it possible for the screws to be used multiple times. The square drive system also facilitates single-hand operation because the screws are more tightly held to the bit.

Figure 2-5. *Square drive.*

TORX

The TORX or hexalobular drive system is designed to increase the contact surface area between the driver and the fastener (Figure 2-6). This system is also designed to not cam-out when turned, which dramatically reduces the need to push on the driver while turning. TORX screws are often found in consumer electronics because the drive system helps to eliminate tool slippage and debris produced due to wear. The TORX drive system also uses standard bit sizes across both metric and standard threaded screws, thus reducing the quantity of tools required.

Figure 2-6. *TORX drive.*

General Tool Use

To use a screwdriver, start by selecting the appropriate-sized drive for the fastener (Figure 2-7). Oversized bits are obvious because they won't fit, but an undersized or incorrect unit system can result in permanent damage to the tool or the fastener. A properly sized bit will fit snugly, with little room for free rotation.

When driving a fastener, fully seat the bit into the head of the fastener and ensure that they are both parallel in alignment. Apply downward pressure toward it to help prevent cam-out.

When removing a fastener, the process can be counterintuitive. Set up the driver as mentioned above and again apply pressure toward the fastener while you counter-rotate the fastener. If it feels like your pressure is preventing the removal of the fastener, reduce the pressure, but not enough to cause cam-out.

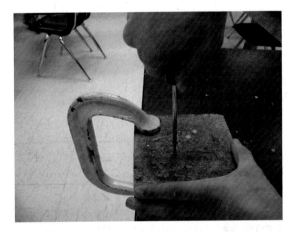

Figure 2-7. *Apply downward pressure on the fastener when driving and removing fasteners with a screwdriver.*

Pliers

Pliers are designed to increase the amount of compressive force exerted by your grip and transfer it to an isolated tip, measured by the systems mechanical advantage. They can be used to turn, cut, bend, and strip a wide range of materials making them indispensable in a Makerspace. Each tool serves a unique purpose, so there are a variety of plier types that are designed to tackle your most challenging tasks.

Properly snipping, stripping, and crimping wire is a process that will ultimately dictate the longevity of the electrical connection. By adhering to the manufacturer's specification for use on designated materials and thicknesses, the tool will continue to function as intended for years. Because these tools feature a cutting edge, they are prone to misuse. Using these pliers on materials that exceed the specification can result in damage to the tool's cutting surface. This permanent damage can inhibit your tool's functionality and can even render the tool unusable.

The common plier features are identified as follows:

Grip

The grip provides the interface between the input, commonly provided by your hand, and the output interface. Grips are mainly made out of steel and can be coated in materials that both increase usability and isolate your hands from steel's electrical conductivity.

Head

The tool head is designed to optimize the transfer of compression from your hand to your work. Depending on the task, heads can be designed to grip, snip, strip, and crimp.

Pivot

The pivot acts as the fulcrum of this lever system. Pivots are either adjustable, comprising a nut/bolt combination, or are fixed and held together with a rivet or similar fastener.

Gripping Pliers

Combination-Jaw

Combination-jaw pliers are designed to handle general purpose tasks (Figure 2-8). This plier type features a slip-joint pivot that allows the jaw opening to be expanded for larger objects. Other combination-jaw plier features include a wire cutter at the root of the head and serrated jaws for added grip.

Figure 2-8. *Combination-jaw pliers.*

Locking

Locking pliers, or vise grips, are designed to dramatically increase mechanical advantage and employ a grip-locking mechanism to retain compression (Figure 2-9). This plier type contains a pressure-adjusting screw that changes the amount of compression required to engage the grip-lock. The mechanism is then released when a trigger located on one handle is squeezed. Additional features include longnose heads, pinch-off jaws, and wire cutters.

Figure 2-9. *Locking pliers.*

Needle-nose

Needle-nose pliers are designed to extend the head into tight areas and bend light gauge wire (Figure 2-10). This plier type is good for working on small tasks because the compressive capability is decreased due to the loss of mechanical advantage and decreased material strength. Additional features include angled heads, wirecutters, and serrated jaws.

Figure 2-10. *Needle-nose pliers.*

Nonmarking

There are times when you need to apply grip to a part while limiting the potential for surface damage (Figure 2-11). Nonmarking pliers are made of soft materials like brass, or use nylon covers to prevent damage while gripping.

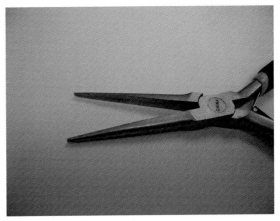

Figure 2-11. *Nonmarking pliers.*

Tongue-and-Groove

Tongue-and-groove pliers function similarly to combination jaw pliers in their gripping capability (Figure 2-12). This plier type employs a sliding top handle and angled head with which you can adjust the jaw opening while not affecting the mechanical advantage.

Figure 2-12. *Tongue-and-groove pliers.*

General Tool Use

To use pliers, start by determining the correct plier type for the application (Figure 2-13). If the set of pliers are too small for the application, they will not transfer enough force to the workpiece, whereas using pliers that are too large for the application will damage the material by delivering too much force.

Most pliers work best at holding rather then wrenching. This pertains to pliers that feature toothed jaws as the teeth embed themselves in the workpiece when compressed. If torque is then applied to the pliers, the teeth will plow across the surface of the workpiece and will result in irreparable damage. If it is absolutely necessary to use pliers as a wrench, you can try wrapping your workpiece with a piece of cloth or consider using a set of pliers that feature a nonmarking jaw.

Figure 2-14. *Sheet metal shears.*

Shears Use

To cut metal with shears, start by outlining the path of the cut onto the surface of the material (Figure 2-15). Multiple shear types will be needed depending on the complexity and curvature of the path. Release the blade-lock mechanism and start cutting along the path. As you cut, the material will be pushed above and below the shears resulting in a slight bend. This bend can be removed by gently hammering the metal back into place. Continue cutting until the path is complete.

Figure 2-13. *If a second wrench is not available to remove a fastener, use a pair of pliers to hold the fastener stationary while twisting only the wrench.*

Crimps, Shears, Snips, and Strips
Shears

Sheet metal shears are used to cut and trim sheet metal up to about 18 gauge in thickness (Figure 2-14). Each pair has a color-coded handle that illustrates the direction in which the tool makes a cut. Shears with a yellow handle are designed to cut in straight lines and light curves; red-handled shears are designed to cut to the left; and green-handled shears are designed to cut to the right.

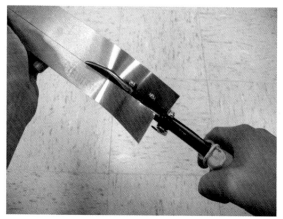

Figure 2-15. *Use the appropriately colored shear that matches the direction of the cut.*

Wire Crimps

The quickest method for attaching wire to a connector is to crimp it (Figure 2-16). Wire crimps are designed to make this electrical connection by physically compressing a metallic connector onto the wire's conductors. There are various types of wire crimp connectors, but most rely on a split cylinder attached to the desired connector type. To properly crimp this connector type, you must use a multistep procedure. Ensure that you choose a pair of crimps that features a crimping die for both insulated and noninsulated connectors.

Figure 2-17. Wire crimps use compression to secure an electrical connector to the conductor and require a multistep crimping process.

Figure 2-16. Wire stripper/crimper.

Wire Snips

Wire snips really are fantastic tools (Figure 2-18). They range in size from small precision snips used to trim the leads off of through-hole components to large hardened steel snips that can cut nails. Depending on the design, wire snips are generally intended to cut through soft steel, copper, and soft metals. To preserve cutting ability, it is recommended that your Makerspace has multiple types of wire snips available and designate them for specific tasks (for example, copper only, plastic only, and so on).

Figure 2-18. Wire snips.

Wire Crimper Use

Start by stripping 1/4 in (Figure 2-17) of insulation from the end of your wire and insert it into the connector. Position the connector in the first crimping die and compress the jaws. Next, position the connector and wire into the second die, align the finger with the connector's split, and compress the jaws. Finally, position the connector and wire back into the first die and rotate it at a 90 degrees. Compress the jaws once again to finalize the connection.

Wire Snip Use

Snips provide a means for cutting wire to length, removing excess wire, and trimming components (Figure 2-19). Place the material into the jaws of the snips and compress the handle to make the cut. Thicker material is easier cut if you position it as close to the jaw's root as possible. Be sure you that you don't twist or torque the snips as you cut, especially when trimming the legs of soldered components. This torquing can cause damage to the tool by misaligning the jaws.

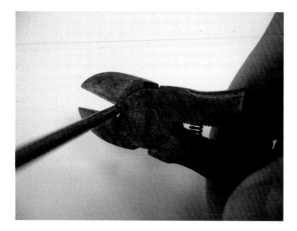

Figure 2-19. *Wire snips produce more cutting power at the root of their jaws.*

Wire Strippers

Wire strippers are designed to remove the outer insulation that surrounds electrical wire (Figure 2-20). These tools contain a series of graduated cutters or an adjustable head, either designed to cut just deep enough to separate the insulation from the copper, without cutting the wire. Some wire strippers are also equipped with a wire cutter, bolt cutter, and plier nose. There are also wire strippers that "automatically" strip the wire by using a clamping jaw mechanism. Although these tools can successfully strip wire, they often do so poorly and can even stretch the wire causing potential internal damage.

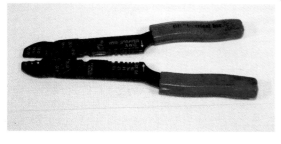

Figure 2-20. *Wire strips.*

Wire Stripper Use

Stripping wire is not a trivial task and often requires a bit of guess and check. Every pair of wire strips features a wire gauge that indicates the size of wire to be stripped. Because the thickness of the insulation around the wire can vary in size, the gauge indicates the conductor thickness not the insulation. Start by positioning your wire in the designated cutter and compress the jaws. This action will force the cutting surface over and around the wire, and if used properly, will cut only the insulation. With the jaws still compressed, carefully push the strippers away from you so as to remove the desired amount of insulation. In general, only 1/8 to 1/4 in (Figure 2-21) of insulation should be removed. If you do not know your wire gauge, you can choose the closest size as long is it is not smaller. Using a cutter that is meant for a smaller gauge will actually cut into your wire, resulting in damage to the strands or core. If this happens, snip off the wire at the cut and try again.

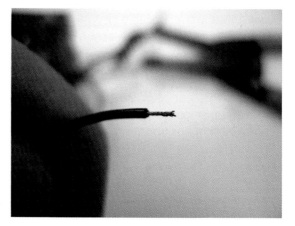

Figure 2-21. *Remove only enough insulation to make a proper connection. This is usually between 1/8 and 1/4 in.*

Wrenches

Wrenches are similar to pliers in their ability to grip a fastener, but are designed to apply even force over the fasteners head without causing damage. Each wrench is designed to turn a specific fastener size, or range of sizes if adjustable, and contain markings that illustrate the intended size.

The common wrench features are identified as follows:

Handle

The handle is designed to transfer the force of your hand through the head and around the fastener. Some handles are designed to offset the position of the head, allowing for easier access to recessed fasteners.

Head

A wrench's head is designed to mate with a similarly sized fastener. Heads can be either open or closed and feature a 2, 6, or 12 point interface so that you can manipulate the wrench at varying angles. Ratcheting heads allow for the handle to completely rotate the head while moving only a couple of degrees.

Wrench Types

Adjustable

You can use an adjustable wrench as a universal open-ended wrench for loosening or tightening fasteners and can be set to any size within the specified range (Figure 2-22). These wrenches tend to loosen while being wrenched and therefore work best when used in conjunction with a fixed-sized wrench. Use the adjustable wrench to hold one side of the fastener while turning the other side with the fixed-sized wrench.

Figure 2-22. *Adjustable wrenches.*

Box

A box wrench is designed to fully enclose the fastener's head, utilizing either a 6- or 12-point opening (Figure 2-23). By enclosing the entire head, the contact area is dramatically increased, which provides better torque transfer as well as prevents damage to the facets. These wrenches are also available with offset heads at typically 15 and 45 degrees, affording easier access into recessed areas.

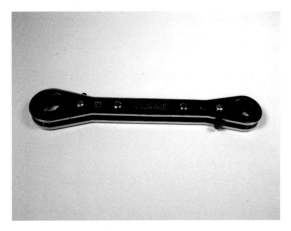

Figure 2-23. *Ratcheting box wrench.*

Open-Ended

An open-ended wrench features a split opening so that you can slide the wrench around the fastener (Figure 2-24). This design is intended for use when access to the entire fastener is obscured. Open-ended wrenches have a tendency to damage the fastener's head because only two points of contact are made with the facets.

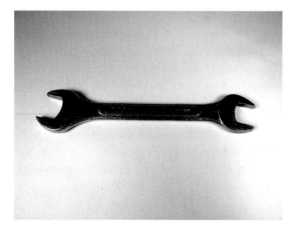

Figure 2-24. *Open-ended wrench.*

Combination

Combination wrenches contain both box and open-ended heads, extending the tool's functionality (Figure 2-25). Some combination wrenches feature offset heads that make it easier to access awkwardly positioned fasteners.

Figure 2-25. *Combination wrench.*

Pipe

You use pipe wrenches specifically to turn metal pipe (Figure 2-26). This wrench features an adjustable pivoting head that grips the outer pipe wall when engaged. Because this tool compresses the pipe when turned, it is not recommended for soft metal and plastic pipes.

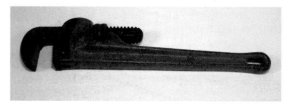

Figure 2-26. *Pipe wrench.*

Socket

Socket wrenches feature a ratcheting head containing a 1/4, 3/8, or 1/2 in square drive mechanism (Figure 2-27). This mechanism interfaces with a wide range of socket sizes, making the socket wrench incredibly versatile. Sockets are attached to the head by depressing the button on the back of the head, which releases the locking mechanism. When in use, the direction of the ratcheting can be changed by sliding the lever located beneath the locking mechanism.

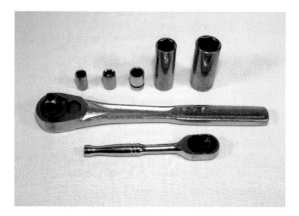

Figure 2-27. *Socket wrench.*

Torque

A torque wrench functions like any other wrench, but includes the ability to tighten a fastener to a specific amount of torque (Figure 2-28). These wrenches typically have a dial mechanism with which you can select the desired torque, which is measured in ft/lbs or N/m. The torque is indicated as the wrench is turned on the indicator or through a light snap, depending on the design.

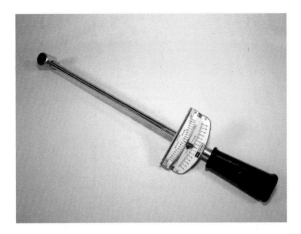

Figure 2-28. *Torque wrench.*

General Tool Use

Wrenches are designed to apply enough torque to either loosen or tighten a fastener. As wrenches can greatly amplify the amount of torque delivered to the fastener, it is important to choose the right tool for the job and use it properly. Start by determining the size of the fastener, and the correct mating wrench. When your wrench is secured on the head of the fastener apply force to the handle in the desired direction until the fastener is either fully secured or removed (Figure 2-29).

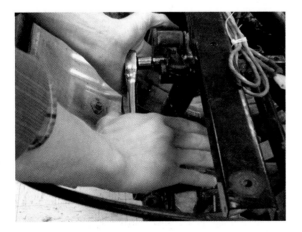

Figure 2-29. *When you are attempting to loosen a stuck fastener and you are in close quarters, never wrap you hand completely around the handle. Rather push with your palm and leave your fingers extended. This greatly*

reduces the risk of unintentionally punching the surrounding area when the fastener breaks free.

Hammers

The hammer is simply a blunt object that is meant to amplify and transfer the force generated by the human arm. Today's hammers feature ergonomic handles and sophisticated head designs that improve this transfer of force while reducing strain on the person wielding it.

The common hammer features are identified as follows:

Handle

The handle is designed to both support the head as well as provide a stable gripping interface. There is a wide range of material types and profiles that are designed to provide the most amount of energy transfer to the workpiece.

Head

The hammer's head is considered the "business end" of the tool. Hammer heads are designed to repeatedly deliver impact forces to the workpiece and often work in combination with ancillary tools that allow for chipping and prying.

Hammer Types
Ball-Peen

Ball-peen hammers are good all-around hammers (Figure 2-30). The head features two tools: a slightly rounded striking face and a rounded peen. The striking face can be used for driving nails, tapping chisels and punches, and freeing stuck components. The rounded peen end is used mainly for light metal forming and setting rivets.

Figure 2-30. *Ball-peen hammer.*

Figure 2-32. *Nonmarring hammer.*

Claw

The claw or carpenter's hammer has the same slightly rounded striking face as the ball-peen hammer but has a claw in lieu of the ball end (Figure 2-31). The claw is designed for removing nails, ripping, and prying.

Sheet Metal

Sheet metal hammers typically come in a set with corresponding anvils (Figure 2-33). These hammers feature a series of heads that are designed to form sheet metal into practically any shape. With a little finesse and practice, forming metal can be easy and fun.

Figure 2-31. *Claw hammer.*

Figure 2-33. *Sheet metal hammer.*

Nonmarring

Nonmarring hammers are designed to provide the blow of a hammer, but feature a head made out of rubber, dense plastic, or wood (Figure 2-32). These hammers are good for removing stuck components that are susceptible to surface damage.

Sledge

Sledge hammers feature a heavy head that is solely designed for large, forceful impacts (Figure 2-34). The handle length can vary from around a foot to several feet, providing for an impressive blow.

Figure 2-34. *Sledge hammer.*

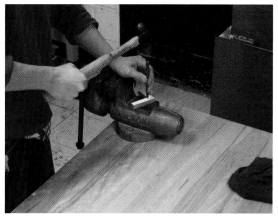

Figure 2-35. *Hold the hammer at the end of the handle and hit the workpiece squarely.*

General Tool Use

Hammers are designed to produce an impactive force as an extension of your arm's strength and ability (Figure 2-35). Because hammers are meant to transfer and amplify this force, it is therefore important to do so correctly. Start by securely gripping the base of the handle. Positioning your hand high on the handle (closer to the hammer head) greatly decreases the hammer's impactive ability and control. Keeping your elbow bent and moving only your forearm, deliver a controlled blow to your work. Generally, it is better to deliver fewer blows with greater intensity than a large quantity of smaller blows. This reduces the risk of damage to the surrounding area and increased arm fatigue. Never swing a hammer so that is passes behind your head. This can result in massive injury to those around you, especially if someone walks behind you or you accidentally let go of the hammer. If you require greater impact force, use a bigger hammer.

Two similar metal hammers should never be hit together as the resulting resonate impact can be great enough to shatter the hammer's head.

Thread Cutting

Threads provide a means for converting rotational motion into linear motion. The resulting mechanical advantage translates into an incredible amount of compressive force to be generated from a relatively small input. Threads can be added to any hole using an appropriately sized tap and to any shaft using an appropriately sized die. Taps and dies are made out of tempered tool steel, which means that they are very strong and resistant to flexing, therefore they are very brittle. If too much torque is applied to the tool while cutting thread the teeth or entire tool will crack and fail.

The common thread-cutting tool features are identified as follows:

Flute

Both taps and dies contain four or more flutes that support the thread-cutting teeth. The flutes also help to eject cut material up and out of the tool as it rotates reducing the chance of binding.

Pitch

The pitch refers to the quantity of threads per inch/mm. Finer pitched threads increase the mechanical advantage but reduce the strength of the threads, whereas coarser threads reduce the mechanical advantage but increase the strength of the threads.

Tap/Die Chart

A chart is included with every tap-and-die set that illustrates the proper hole diameter for each tap and the proper shaft diameter for each die. It is important to match the diameters of the material to the correct tap and die to ensure proper cutting and prolonged tool life.

Thread-Cutting Types

Tap

Taps are cutting tools that are designed to cut threads into the walls of an appropriately sized hole (Figure 2-36). Taper taps are the most common tap type and feature tapered flutes. The taper facilitates aligning and starting the tool. Bottoming taps are designed to cut the complete thread along the length of the tap. These taps are difficult to start, but you can use them to complete the threads created by a tapered tap.

Figure 2-36. *Thread-cutting tap.*

Tap Use

To properly tap a hole, start by referencing the drill/tap chart and drill an appropriately sized hole into your workpiece (Figure 2-37). Apply a small amount of cutting oil to the hole and secure the tap into the holder. Insert and align the tap so that it is parallel with the walls of the hole. Applying a small amount of downforce, carefully rotate the tap one complete turn. This will start the thread action, at which point, you can continue making threads along the length of the hole. Counter rotate the tap 90 to 180 degrees until a small pop is felt, indicating the separation of the chips. Rotate the tool 180 degrees past the previous point, back it off 90 to 180 degrees and continue the process until threading is complete (Figure 2-38). If you feel more resistance then usual, stop tapping. This indicates that you have reached the bottom of the hole or the material is too tough to tap.

Figure 2-37. *Ensure that you use a properly sized hole is used for taping. This reduces the chance of the tap breaking off or forming poor-quality threads.*

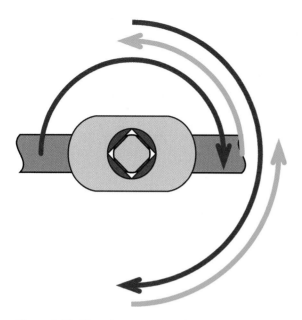

Figure 2-38. *Thread-tapping procedure.*

You might encounter a common problem when tapping small holes in aluminum. Because aluminum is a soft metal, it has a tendency to bind in the tap's thread-cutting teeth. Clear the chips more frequently when tapping aluminum to prevent the tap from jamming. Trying to remove a tap that has broken off in the hole while tapping aluminum is virtually impossible.

Die

Dies are cutting tools that are designed to cut threads onto the outside of a round shaft (Figure 2-39). The die's flutes are tapered on one side and are intended to assist in starting the thread cutting process. Once the desired amount of threads are cut, the die can be flipped around to cut the tapered threads to the proper depth.

Figure 2-39. *Thread-cutting die.*

Die Use

To properly thread a shaft, start by referencing the diameter/die chart to determine the proper die size for the diameter of your workpiece (Figure 2-40). Follow the same procedure as tapping to start and cut the threads (Figure 2-41).

Figure 2-40. *Make sure a properly sized shaft is used for thread cutting. This reduces the chance of the die's teeth breaking off or forming poor quality threads.*

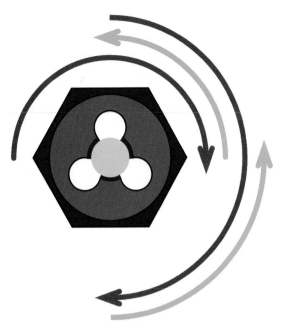

Figure 2-41. *Thread Cutting Procedure.*

Saws

Saws are available in a wide range of configurations and blade types, making it possible for you to cut materials ranging from soft woods to carbon steel. With a little patience and perseverance, no task is too big for a saw.

Common saw features are identified as follows:

Blade

Saw blades are designed to remove a specific type of material efficiently and with the least amount of effort possible. The blade's teeth are typically angled in one direction, allowing for the saw to make a cut and then clear the chips during recoil. The blade type dictates the type of materials that it can cut, whereas the number of teeth determine how aggressively it cuts. When cutting dense materials, cutting oil should be used to cool the blade and prolong its life.

Handle

The saws handle forms the interface between the human hand and the cutting blade. Handles are commonly made out of wood or plastic and are designed to provide

a comfortable and controlled grip while in use.

Kerf

The blade's kerf refers to the final width of the cut. Typically, wood blades are designed to remove more material while cutting, producting a wider kerf, whereas metal blades produce a kerf that is much more narrow.

Saw Types
Coping

Coping saws utilize a thin blade that can be adjusted at 90-degree increments. It is designed to cut through thin woods and plastics (Figure 2-42). Due to its thin blade design and deep throat, the coping saw is optimal for producing curved, intricate cuts as well as making holes without a lead-in cut.

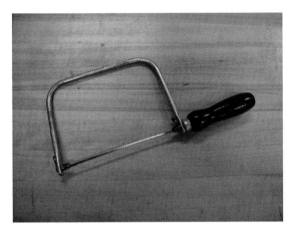

Figure 2-42. *Coping saw.*

Hack

Hack saws are similar to coping saws in design, but provide a longer blade length and higher rigidity (Figure 2-43). This saw is designed for cutting mild steel, wood, plastic, and other soft metals, depending on blade type. The blade is also adjustable in both angle and direction, which allows for holes to be cut in material without a lead-in cut.

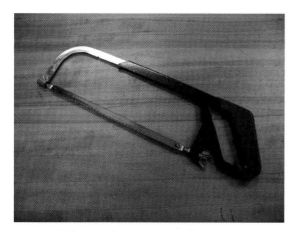

Figure 2-43. *Hack saw.*

Figure 2-45. *Razor saw.*

Handsaw

The handsaw, or rip saw, comprises a short handle connected to a long blade that is used for cutting wood and plastic (Figure 2-44). Due to it's wide blade design, handsaws are optimal for cutting long straight lines.

Backsaw

Backsaws are straight-bladed handsaws that are commonly used to cut tenons and dovetails in wood. Each saw features a reinforced backpiece that helps keep the blade from warping while in use.

General Saw Use

Saws are designed with many teeth that make small cuts as they pass through a material. Depending on the saw's design, the teeth will either be angled away from you, perpendicular to the blade, or toward you. Teeth that are angled away from you will make a cut when the saw is pushed across the material. Teeth that are angled perpendicular to the blade will make less aggressive cuts while being both pushed and pulled. Teeth that are angled toward you will make a cut when the blade is pulled across the material. Always begin sawing by properly securing your material to the workbench in order to prevent it from unintentionally shifting. Make a guide cut in your material by lightly pushing the saw's blade across the surface. The resulting channel will act as a guide for the blade when you begin the actual sawing process.

Cutting oil (Figure 2-46) helps to keep the blade cool, makes the cutting process easier and prolongs the life of the blade.

Figure 2-44. *Handsaw.*

Razor

Razor saws are designed for cutting small pieces of wood, plastic, or soft metal, with a very narrow kerf (Figure 2-45). This saw is an excellent substitute for a razor blade and often produces less deformation to soft materials like balsa wood.

Figure 2-46. *Use cutting oil whenever you saw metal.*

Continue to make cuts into your material while applying a slight downward force during cutting strokes and remove the pressure during recoil. Be sure that you utilize the full extent of the saw blade rather then making short, quick cuts. This not only helps reduce fatigue while cutting, but ensures even wear to the blade.

Sanding and Filing

Sanding and filing are subtractive processes that are designed to remove only a small amount of material in a controlled manner. These tools are commonly referred to as finishing tools because they are mostly used during the final stages of a project.

The common sandpaper and file features are identified as follows:

Grade

Coarseness is based on the quantity of abrasives on the tools working surface and ranges from ultra-fine to coarse. Sandpaper and files can be used to remove small amounts of material over wide areas as well as polish material to a mirror finish.

File Types
Half-Round

Half-round files provide both flat and convex sides so that the tool can work on both rounded and flat surfaces (Figure 2-47). It can even be used to make narrow notches using the edges.

Figure 2-47. *Half-round file.*

Rectangular

Rectangular files are general purpose files that are designed to debur, shape, and sharpen most materials (Figure 2-48). This file features three cutting surfaces and one plain edge.

Figure 2-48. *Rectangular file.*

Round

Round files are designed to access cut holes and other hard-to-reach areas (Figure 2-49). Because the entirety of the file is round, it is not recommended for use when flattening a surface or edge.

Figure 2-49. *Round file.*

General File Use

Unlike a saw, files are able to cut material in both directions (Figure 2-50). To make a smooth and level cut, start by securing your workpiece to the worktable using a vise or clamp. Grip the file on the handle and use your free hand to guide the file over the workpiece while pushing. Lift up the file and repeat the cut until the desired surface finish is achieved.

Figure 2-50. *Use one hand to push the handle and the other to guide the file.*

When cutting soft metals, make sure that the files teeth do not become clogged with shavings. A file brush can be used to remove stuck material and renew the file's cutting ability.

Sandpaper Types
Belt

Sandpaper belts are typically used with power tools as a means of continuously drawing the paper over the workpiece (Figure 2-51). These belts are available in the same grits as sandpaper sheets and sold by the belt length.

Figure 2-51. *Sanding belt.*

Drum

Sanding drums are typically used with power tools and are designed to sand flat and irregular surfaces (Figure 2-52). The drums function similarly to a sanding belt in as much as the sanding surface is continuously rotated over the workpiece.

Figure 2-52. *Sanding drum.*

Flap Wheel

Flap wheel sanders are typically used with power tools and are designed to provide gouge-free sanding (Figure 2-53). The flapping action and flexibility of the paper allows for flap wheels to be used in obscure and complex areas that normal sanding methods could not reach.

Figure 2-53. *Flap wheel sander.*

Sheet

You will find sandpaper most commonly available as sheets, and you can use it by simply rubbing the paper onto the workpiece, or secured into a sanding block for a more controlled use (Figure 2-54). Finer grit

sandpaper can even be used when wet to promote the cutting and polishing action.

Figure 2-54. *Sandpaper sheet.*

Sandpaper Use

Sandpaper only removes a small amount of material as compared to other cutting tools and thus requires patience. Start by determining the type and grade of the sandpaper required. Sanding typically starts with a coarse grade sandpaper, gradually moving to a finer grade. Working along the grain, or in a circular pattern (Figure 2-55), apply moderate pressure to the paper as you sand. Continue this process until the desired surface finish is achieved.

Figure 2-55. *Sand in a circular pattern to create an evenly sanded surface.*

Punches

Punches are designed to transfer energy from a hammer or lever to a focused point in a controlled manner. When used with a hammer, punches have the ability to mark material, drive nails, and transfer locations. Lever punches transfer their energy through the action of a lever or screw, gradually amplifying the input force.

The common punch features are identified as follows:

Point Diameter

The point diameter specifies the diameter of the tip of the punch. For center and transfer punches, the tip diameter is very small, whereas nail and pin punches feature a wider flat point.

Punch Diameter

The punch diameter illustrates the actual diameter of the punch and is applicable to sheet metal and roller chain punches.

Punch Types

Center Punch

You use a center punch in conjunction with a hammer to create a small indentation into the workpiece (Figure 2-56). This indentation acts as the starting point for drilling and helps to prevent the bit from wandering as it turns. This punch type has a tapered steel shaft with a knurled handle for added grip.

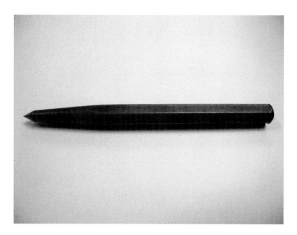

Figure 2-56. *Center punch.*

Center Punch Use

To use a center punch, start by marking the location of the desired hole using a marker or scribe (Figure 2-57). Position the point of the center punch at the desired location and hit the top of the punch only once with a hammer. Multiple hits will result in the punch jumping position and marking an undesired location.

Figure 2-57. *Hit a center punch only once to prevent it from jumping.*

Roller Chain Punch

You purchase roller chains by link quantity. They are available as a circular loop or an unconnected chain (Figure 2-58). When a chain needs to be completed or broken, the pin that holds the links together needs to be removed. The roller chain pin punch provides the necessary support and action to properly push the pin out of the chain without torquing the links. When the desired length is obtained, the links can be rejoined with a master-link or the punch can be used to reset the original pin.

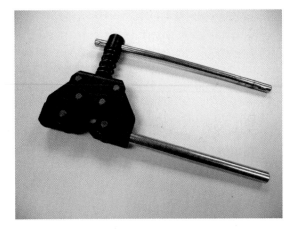

Figure 2-58. *Roller chain punch.*

Roller Chain Punch Use

To use a roller chain punch, start by opening the punch's jaws and clamp them over the link pin that is to be removed or installed (Figure 2-59). Slowly rotate the handle until the pin is punched out of the chain. Back off the punch and remove from the chain when completed.

Figure 2-59. *Be carful to align the punch over the pin because the punch can produce enough force to bend the links if not properly aligned.*

Sheet Metal Punch

It is a pain to drill a hole into sheet metal (Figure 2-60). The drill bit tends to wander as you begin to rotate it, and the result is often a noncircular deformed mess that tends to bind to the bit as you drill. Sheet metal punches are designed to quickly produce clean and accurate holes in sheet metal with little effort. The tool uses an interchangeable punch and die set that is connected to a pressing mechanism that forces the punch into the die when compressed. When the sheet metal is fed into the opening and the handle is compressed, the punch and die mate and punch out the desired hole. Much like a paper hole punch.

Figure 2-60. *Sheet metal punch.*

Sheet Metal Hole Punch Use

Start by selecting the correct punch and die combination for the desired hole size (Figure 2-61). Open the jaws of the punch and slide it over the sheet metal. Position the punch in the desired location and compress the handle until the punch passes all of the way through the material. Open the jaws and remove the punch when complete.

Figure 2-61. *Sheet metal punches should be used in lieu of drilling as they are less likely to distort the metal.*

Transfer Punch

A transfer punch is very similar to a center punch in design but features a nontapered shaft (Figure 2-62). This punch is designed to transfer the location of pre-made threaded or drilled hole to a blank piece of material.

Figure 2-62. *Transfer punch.*

Transfer Punch Use

Start by determining the size of the hole that is to be transferred and select a similarly sized transfer punch (Figure 2-63). Stack two pieces of material together, one with the hole to transfer and a blank, and secure onto the work surface. Pass the punch through the hole and tap once with a hammer to transfer the mark. These punches are also designed to be slightly undersized to allow the punch to pass easily through the transfer hole.

Figure 2-63. *Choose a transfer punch that matches the size of the hole. This will ensure a correctly transferred center.*

Power Tools

There are times when the force generated by the human body to move a tool is not enough for the task. The purpose of a power tool is to amplify and offset the work done by the body to work done by a machine. Power tools typically utilize an external power source (although there are many that use gas-powered engines), either from an AC electrical outlet or a battery pack, to rotate an electric motor whose torque is amplified via a complex gearbox.

Drills

Choosing the correct drill bit is not a trivial task. They are used by both portable and standing power tools, with the selection being dependent on the material, hole type, diameter, finish, and depth. All of these parameters help to determine a drill bit that will not only produce the desired hole, but will do so time and time again. Most of the materials drilled in a Makerspace are usually wood, plastic, and metal.

Every material requires a unique drill bit spinning at a specific RPM in order to ensure proper cutting. This RPM is determined by calculating surface velocity, as measured in Surface Feet Per Minute (SFM) and Surface Meters Per Minute (SMM).

```
RPM = (SFM × 3.82) / Drill Diameter (in)
RPM = (SMM × 318.057) / Drill Diameter (mm)
```

High-Speed Drill Bit Speed Reference		
Material Type	SFM	SMM
Aluminum	250	75
Brass	225	70
Soft cast iron	125	40
Low-carbon steel	100	30
High-carbon steel	75	25
Stainless steel	50	15
Plastic/rubber	150	45
Wood	125	40

Figure 2-64. *When properly used, a drill bit will eject the cut material in long spirals.*

As you drill, you should see material ejecting from the flutes of the drill bit in the form of a long corkscrew or clean chips. This is a good indication that you have selected the proper drill bit, rotating at the correct speed and plunging at the correct rate. Driving a drill bit that is spinning too slow can cause the bit to bind, potentially causing damage to the material or the operator. Driving it too fast can cause the bit to overheat, po-

tentially damaging your workpiece or even breaking the bit.

The common drill bit features are identified as follows:

Bit Diameter

> The bit diameter refers to the actual cutting diameter of the drill bit.

Flute

> The flute is the channel located beneath the lip. It extends from the middle of the point to the top of the shank. It's purpose is to transfer chips or cut material from the tip and eject it from the cutting area as the bit penetrates the workpiece.

Lip

> The lip is a drill bit's second cutting surface that extends from the edge of the point to the top of the shank. This feature determines cutting speed and aggressiveness of the cut.

Point

> The drill bit point is the most important part of the drill bit because it dictates the type of material to be cut. Drill bit points vary depending on the material type and desired cutting method. Regardless of type, the point is designed to center the bit and perpetuate the cutting action through the entire process.

Shank

> The drill bit shank is responsible for connecting the cutting surfaces of the bit to the drill's chuck. Depending on the drill bit type, shanks can contain flats which help prevent the chuck from spinning around a stopped bit if it is not properly tightened.

Shank Diameter

> The shank diameter refers to the actual diameter of the drill bit. As drill bits increase in size, the shank diameter might not match the diameter of the bit. This allows for larger bits to be used in smaller chucks. Typical drill chuck sizes are 3/8 in and 1/2 in.

Drill Bit Types
Brad Point

The brad point bit features an extended point that helps center the bit when cutting and prevents wandering (Figure 2-65). These bits produce accurate smooth holes and work well when drilling across the wood grain.

Figure 2-65. *Brad point drill bit.*

Forstner

Forstner bits are designed to cut a straight and smooth finished hole (Figure 2-66). You can purchase these bits with and without a center point, making it possible to drill a variety of holes and wood types.

Figure 2-66. *Forstner drill bit.*

General Purpose

General purpose, high-speed drill bits are designed to cut metal and some nonmetals such as plastic and fiberglass (Figure 2-67). These bits typically feature a 135-degree point angle that produces small chips, and a split point which keeps the bit centered. You can also purchase general purpose drill bits that have surface coatings like titanium nitride that help to maintain cutting performance.

Figure 2-67. *General-purpose drill bit.*

Masonry

Masonry bits are designed to withstand both impact and torque forces experienced when drilling masonry and rock (Figure 2-68). These bits feature a reinforced point that both cuts and shatters the material as it drills.

Figure 2-68. *Masonry drill bit.*

Plastic

Although general purpose drill bits can be used to cut plastic, the heat produced during drilling can distort or crack the material (Figure 2-69). Plastic bits should be used when drilling plastics, such as acrylic, nylon, and PVC, which are soft and distort easily. Plastic cutting drill bits feature a 60-degree point angle that eliminates edge chipping, a problem that can produce small cracks on the walls of the hole.

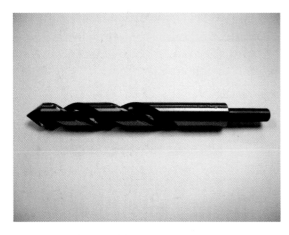

Figure 2-69. *Plastic drill bit.*

Spade

A spade bit features a flat blade and grooved point that can quickly cut through hard and soft wood (Figure 2-70). Due to its flat design, it is often difficult to drill a perfectly straight hole.

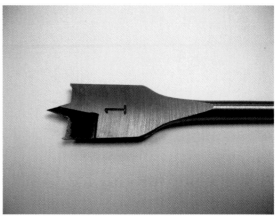

Figure 2-70. *Spade drill bit.*

Drilling Metal

Drilling metal is a bit harder than drilling wood, both literally and figuratively. Start by identifying the type of metal and hole size in order to determine the appropriate cutting speed. You will also need to use cutting oil, a center punch, and a hammer. Mark the location of the hole by using the center punch and secure the bit in the chuck. Add a small amount of cutting oil to the surface of the workpiece so that it covers the punch mark. Ensure that the drill is spinning in the correct direction and align the drill and drill bit so that they are perpendicular to your workpiece. Spin up your drill to the correct RPM and slowly plunge the bit into the workpiece. You should see chips ejecting from the flutes of the bit. Drill about 1/4 in. into your workpiece and retract the drill. Add oil to the hole (Figure 2-71) and continue drilling for another 1/4 in. Continue this process until the desired depth has been achieved. You can use a countersink bit to clean up any burrs around a hole (Figure 2-72).

Figure 2-71. *When drilling metal, use plenty of cutting oil and clear the chips often to prevent the bit from overheating.*

Figure 2-72. *Drill centers can be used to clean burs from smaller diameter holes.*

Drilling Wood and Plastic

Drilling wood, although easier than drilling metal, can be a bit of a challenge (Figure 2-73). As wood is a natural material, it exhibits inconsistent densities that can add complexity to the procedure.

Start by identifying the type of wood and hole size in order to determine the appropriate cutting speed. After you have chosen the appropriate bit, secure the bit in the chuck and ensure that the drill is spinning in the correct direction. Align the drill and drill bit so that they are perpendicular to your workpiece. Spin up the drill to the correct RPM and slowly plunge the bit into the workpiece. You should see chips ejecting from the flutes of the bit. If you are drilling a hole that is deeper then 1 in, periodically retract the drill bit to ensure that the chips have cleared from the flutes. This will help to prevent your bit from overheating and maintain a smooth wall finish. When you have completed drilling the hole, carefully retract the drill until the bit is completely free before turning off the tool.

Figure 2-73. *When drilling plastic, work slowly and clear the chips often to prevent cracking.*

Place a piece of masking tape or clamp a piece of scrap wood to the back of your workpiece prior to drilling. This will help prevent wood from splintering as the drill bit passes through the back of the material.

Portable

Portable power tools are designed to amplify and control the tool's output power by connecting the tool to a motor through a gearbox and linkage system. Although these tools are portable, they are capable of delivering a large amount of force. The tool gets its power from either a connection to an electrical outlet (corded), or from a battery pack (cordless). There are some that even use small gas-powered engines, although you likely would not be using them in a Makerspace so we'll focus primarily on electric-powered tools. When the trigger or switch is activated, power from the source flows into the motor and through a series of mechanical linkages and gears until it is transferred to the tool holder. The power source for electric devices delivers a considerable amount of energy while in operation and therefore requires a good electrical connection and battery capacity. Today's battery-powered tools are powered by either nickel cadmium (NiCd), nickel metal hydride (NiMH), lithium polymer (LiPo), or lithium iron phosphate ($LiFePO_4$) battery chemistries, and therefore vary considerably in power density. As with any tool, care must be taken while operating the tool in order to prevent unintended damage and injury.

The common portable power tool features are identified as follows:

Body

The body is designed to provide a gripping interface to the tool as well as supporting all of the internal components. Design features include ergonomic grips and work lights that help improve usability.

Head

The head is designed to connect the motor and drive system to the tool. Drills feature a rotating chuck that grips the bit, saws feature a blade clamp that secures the blade while cutting and sanders feature a padded surface that holds the paper in place while sanding.

Power Source

Portable power tools that utilize power from an outlet feature a cord extending from the base of the tool. This cord should be plugged directly into the outlet when in use and removed promptly when completed. Power tools that utilize batteries feature an interlock mechanism that couples the battery to the tool's grip. The battery is installed and removed by depressing the release, allowing the battery to slide into or out of place. Be sure to use the correct charger to charge the battery because incorrectly charging certain types of battery chemistries can result in fire. The most common battery chemistries and their relative capacities are illustrated in Table 2-2.

Table 2-2. Portable power tool battery chemistry comparison

Chemistry	Energy Density (Wh/kg)	Charge Cycles	Cost
NiCd	40	1000	$
NiMh	90	500	$$
LiPo	200	500	$$$
$LiFePO_4$	120	2000	$$$$

Portable Power Tool Types
Circular Saw

A circular saw uses a circular blade that is designed to cut mostly wood and plastic materials (Figure 2-74). The saw also features a blade guard that is designed to prevent unintended contact with the blade as well as protecting it from your work surface if the tool is set down while the blade is still spinning. Circular saws are excellent tools for creating long, straight cuts. Curving the tool while cutting can result in the blade binding or kicking out.

Figure 2-74. *Circular saw.*

Circular Saw Use

To use a circular saw, start by ensuring that the appropriate blade is secured in the blade holder and that the teeth are angled in the direction of rotation (Figure 2-75). Make sure that the tool is not in line with your body as you cut to prevent injury if the tool kicks back. Gently rest the saw so that the base is flat against the workpiece and the blade can spin free. Turn on the saw and wait until the blade reaches full speed, then begin to make your cut. Move slowly and carefully along the material until the cut is complete. DO NOT stop the saw while cutting and try to start again while the blade is still embedded in the workpiece. This will cause the saw to kick or stall the motor. Instead, stop the saw, wait until the blade stops spinning, carefully remove it from the workpiece, and start from the beginning. Ensure that the blade guard fully covers the blade when cutting is complete in order to prevent unintended injury.

Figure 2-75. *Make certain that the saw is level against the workpiece and not in line with your body.*

Drill

The drill is designed to secure a bit—whether it be a drill bit or a driver—in its chuck and provide a means for continuously rotating the bit during operation (Figure 2-76). All drills feature a rotation switch to control the direction of rotation. Drills also have the ability to change both the speed and the torque applied to the drill bit or driver. This is accomplished both with a speed switch and through a variable speed trigger.

Figure 2-76. *Drill.*

Drill Use

To use a portable drill, start by configuring the drill's gearbox to rotate at the appropriate RPM and direction for the application. Select the desired bit or driver for the application and secure it into the chuck. If the chuck is keyless, the bit is secured by rotating the outer body of the chuck until tight. If the chuck requires a key, place the tip of the key into the matching hole in the chuck and rotate until tight. When drilling, mark the location of the hole using a center punch. The resulting mark will help to align the bit. Place the bit or driver over the workpiece so that it is perpendicular to the surface. This can be accomplished by sighting down the body of the drill and approximating its position.

When driving into the workpiece, ensure that the driver is spinning clockwise and apply steady press to the drill in the direction of the hole you're drilling (Figure 2-77). Slowly squeeze the trigger until the desired speed is achieved and the process is complete.

When removing a fastener, ensure that the driver spins counterclockwise and apply steady pressure to the drill. This will help prevent the tool from camming out. Slowly squeeze the trigger until the the fastener is completely removed.

Figure 2-77. *Apply a small amount of downforce while drilling and make sure the drill is perpendicular to the workpiece.*

Hammer Drill

A hammer drill functions similarly to a standard drill, but in addition to providing a rotating force, it also applies a reciprocating impact force to the tip of the drill bit (Figure 2-78). This reciprocating action chips away material while drilling and is typically used when drilling masonry.

Figure 2-78. *Hammer drill.*

Hammer Drill Use

To use a portable hammer drill, start by configuring the drill's gearbox to rotate at the appropriate RPM for the application and ensure that only tools designed for hammer applications are used (Figure 2-79). Secure the bit into the chuck and position the drill perpendicular to the work. Gently squeeze the trigger until the job is complete.

Figure 2-79. *The pounding action of the hammer drill helps to break apart hard masonry.*

Impact Driver

Internal to the driver is a swinging mass that acts to impact the tool when rotated (Figure 2-80). This impact provides a spike in the driver's torque and allows for bolts and screws to be driven and removed more easily. You must take care when using an impact driver because the large amount of torque delivered can actually shear fasteners in half.

Figure 2-80. *Impact driver.*

Impact Driver Use

To use an impact driver, ensure that only tools designed for impact driving are used (Figure 2-81). Failure to use the correct bit or driver will result in mechanical failure. Secure the bit or driver into the chuck and position the driver perpendicular to the work. Gently squeeze the trigger until the desire amount of force is applied and the job is complete.

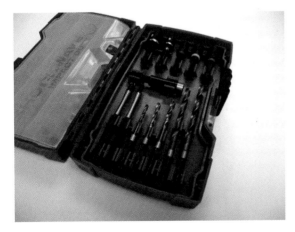

Figure 2-81. *Use only bits and sockets that are designed for impact driving; normal bits will crack under the load.*

Jig Saw

A jig saw, also known as a scroll saw, features a reciprocating interchangeable blade and is great for light cutting (Figure 2-82). Depending on the blade type, these saws are great at cutting tight curves in metal, plastic, and wood. Jig saws typically have an adjustable speed switch for slow and fast cutting that you should set with respect to material type.

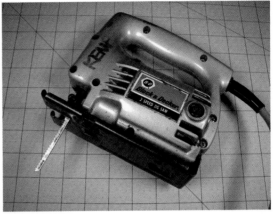

Figure 2-82. *Jig saw.*

Jig Saw Use

To use a jig saw, start by securing the appropriate blade into the tool holder (Figure 2-83). Position the tool in line with the workpiece and ensure that it sits flat against the surface. Without the blade touching the material, pull the trigger and allow the tool to reach the appropriate cutting speed, and then slowly push the tool along the desired path. Tools with blades do not require a lot of force to complete the cut. Applying too much force while cutting will slow down the blade, resulting in it overheating and binding. Continue cutting until the job is complete.

Figure 2-83. *A jig saw can be used to make a cutout within a piece of material. This is accomplished by outlining the desired shape and then drilling a hole within the shape that is larger then the saw's blade. The saw can then be guided around the outline, leaving the outside edges of the material untouched.*

Reciprocating Saw

Reciprocating saws feature a reciprocating blade that is good for making rough cuts (Figure 2-84). These saws provide a lot of power to the blade and can cut through tough materials very quickly. You must take care to ensure that the blade is long enough to make the cut; if not, they have a tendency to kick back when in use and can cause unintended injury.

Figure 2-84. *Reciprocating saw.*

Reciprocating Saw Use

To use a reciprocating saw, start by securing the appropriate blade into the tool holder (Figure 2-85). Position the saw so that it is not in line with your body. This helps to prevent injury if the saw kicks back during operation. Follow the same procedure for starting and operating the scroll saw to achieve a proper cut.

Figure 2-85. *Ensure that the saw's foot is pressed firmly against the material and the saw is not in line with your body.*

Sander

Sanders feature an orbital or belt-driven sanding surface that dramatically decreases the time required to smooth and shape a project (Figure 2-86). These sanding surfaces can be interchanged with different grit densities depending on the type of sanding you need to do.

Figure 2-86. *Power sander.*

Figure 2-87. *Power sanders have the ability to remove a lot of material quickly. Take care to move the tool constantly to preventing pitting the workpiece.*

Sander Use

To use a sander, start by selecting the desired sandpaper coarseness and secure the paper into the holder (Figure 2-87). For orbital sanders, there are two friction clamps that secure the paper onto either side of the holder. For belt sanders, you actuate a lever that retracts the rollers toward each other, allowing for the sanding belt to be installed and removed. With the belt in place, you expand the rollers to secure the paper. Grip the tool tightly and raise it from the work surface. Gently pull the trigger until the tool reaches full operating speed. Carefully lower the tool onto the workpiece and apply a small amount of downforce. Move the tool in a circular path to ensure even sanding and continue until the job is complete.

Sewing Machine

Sewing machines have been used for Making for centuries, and should feel right at home next to any other power tool (Figure 2-88). This mechanical device facilitates the rapid creation of stitches that are used to tack, join, and hem material much quicker and more accurately than if done by hand. In any Makerspace, a sewing machine is the go-to tool for any type of thin material work, ranging from simple fabric crafts to soft circuits.

Although there are a wide range of sewing machine models and types, they all rely the same basic set of features. Each machine has a means for transferring thread through a series of tensioners and through the sewing needle itself. Below the feed dogs, thread from the bobbin is fed through the hook, which, through a series of complicated motions, weaves itself around the needle's thread, producing a stitch. The length, width, and style of stitch are determined by adjusting a series of dials to achieve the desired design (Figure 2-89).

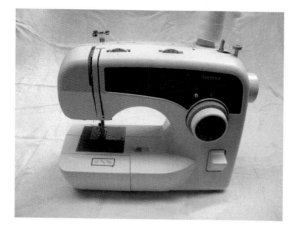

Figure 2-88. *Sewing machine.*

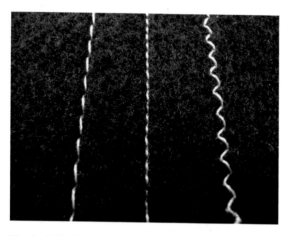

Figure 2-89. *The most common stitch types are the basting stitch, used as a temporary stitch; the straight stitch, which is universal; the zig-zag stitch, used to finish edges with stretch fabrics and to attach patches.*

Laser engravers make cutting fabric easy. Simply draw your pattern in CAD and place as many layers of fabric as needed into the machine. Tack down the fabric with masking tape to prevent it from blowing away when the ventilation system is turned on!

Sewing Machine Use

You use a sewing machine (Figure 2-90) by first preparing a bobbin with the desired thread type (Figure 2-91). Mount the thread

Needle Reference

Needles range in size from 8 to 18 (in the United States) and various point types that dictate the type and thickness of material through which the needle can pass. Use Table 2-3 to determine the best needle size for the job.

Table 2-3. Sewing machine needle reference

Category	Size	Material Types
Delicate	9	chiffon, lace, silk
Lightweight	11	batiste, taffeta, velvet
Medium-weight	14	gingham, linen, flannel, muslin, wool
Medium/heavy-weight	16	drapery, tweed
Heavy-weight	18	canvas, denim, ticking, upholstery

spool on the spool pin and thread about an inch of thread through the side of the bobbin. Place the bobbin onto the bobbin winder and slide over the stopper. Power on the machine, and gently press the foot control until the bobbin begins to wind. When the bobbin is full, the stopper will activate and discontinue the winding process. Remove the bobbin from the winder, trim the thread, and insert into the bobbin case as directed by your particular machine.

Following the diagram, thread your machine by directing the thread around the tensioner, onto the take-up lever, back up to the thread guide, and down to the needle. The thread should pass through the needle from front to back. Lift up the presser foot and while holding the end of the thread in one hand, manually turn the handwheel until the needle enters the bobbin case and pulls up the bobbin thread. Pull the two threads toward the back of the machine and trim them so that about 4 in of thread is exposed. The machine is now ready for operation.

Set the desired stitch type, length, and width, and set your fabric below the presser foot. Lower the presser foot and start your stitching by gently pressing the foot control until you have created a stitch that is approximately 1/2 in long. Press and hold the reverse stitch button and reverse stitch back to the beginning. This process helps to prevent the thread from unraveling, and you are ready to complete the run. After you have sewn to the desire length, repeat the reverse stitch procedure, lift the presser foot, and pull away your fabric. Trim the attached thread and your stitch is complete!

Figure 2-90. *Use your hands to gently guide the fabric. Just watch out for your fingers!*

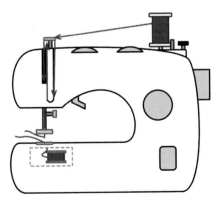

Figure 2-91. *Sewing machine threading procedure.*

Standing

Standing power tools are designed to deliver a large amount of cutting energy in a stable and controlled manner. These tools are typically made out of steel and cast iron components, which make them ideal for jobs that require accuracy and repetition. Smaller power tools can be installed by mechanically securing them to a work surface, while larger tools are typically equipped with a stand.

The common standing power tool features are identified as follows:

Base

All standing power tools feature a robust base that is designed to stabilize the tool while in operation. For smaller power tools, the base can be clamped or bolted to the work surface. Larger tools feature a base that is often heavy enough to prevent unintended movement.

Drive Pulleys

A typical standing power tool utilizes a large electric motor that is coupled to the spindle or blade through a pair of belted pulleys. These pulleys allow for the machine to be configured to rotate at different speeds, allowing for proper cutting of multiple material types. To change the input/output ratio, loosen the tension on the belt and reposition it onto the appropriate pulleys.

Table

The table provides a flat and adjustable surface for moving and securing the workpiece. On the drill press, the height and angle of the table relative to the spindle can be adjusted. On the band saw, the angle of the table can be adjusted relative to the blade. Band saws also feature a channel down the table that is used with a miter gauge, allowing for angled cuts.

Standing Types
Band Saw

The band saw provides a continuously rotating blade that is designed to make long complex cuts through wood, plastic, and metal materials (Figure 2-92). These tools are available in both vertical orientation, for cutting by hand, and horizontal, for cutting stock material.

Figure 2-92. *Band saw.*

Band Saw Use

To use the band saw, start by configuring the blade speed for the type of material to be cut. When this is complete, place the workpiece onto the table and position it next to the the blade. Lower the blade guide until it sits 1/4 in above the surface of the material (Figure 2-93). This helps prevent the blade from wobbling while cutting and will produce better results. Reposition the workpiece away from the blade and power on the machine. Allow the blade to reach full speed and slowly push the workpiece into the blade. As with all cutting tools, this process does not require a large amount of force. The blade is designed to remove small amounts of material at a time, and applying too much force when cutting will slow down the blade, causing it to overheat and bind.

There are two methods for cutting a curve using a band saw (Figure 2-94). The first method is to use a thin blade meant for scroll work. This type of blade permits the material to rotate freely without binding. Alternatively, relief cuts can be used to cut a curve with a wide blade. These cuts help to relieve binding pressure encountered when the blade is torqued. Carefully cut multiple lines perpendicular to the curve at 1/4 in intervals until the curve is fully relieved. Cut along the path of the curve and the relief cuts will allow the blade to operate without binding.

Figure 2-93. *Lower the blade guide until it is just above the workpiece.*

Drill Press

A standing drill press is a great universal tool (Figure 2-95). Not only does it provide a stable drilling platform and means for controlling rotational and plunge speed, but it can serve as a press for bearings and pins. Standing drill presses are available as tabletop or free standing, depending on the power required. Drill presses can also double as a drum sander when the appropriate bit is used and the quill is locked.

Figure 2-94. *Cutting a curve.*

Figure 2-95. *Drill press.*

Drill Press Use

To use a standing drill press, start configuring the spindle speed for the type of material to be cut (Figure 2-96). When this is complete, secure the desired bit into the chuck using the chuck key. Do not forget to remove the

chuck key from the chuck! This is a common occurrence and results in the chuck key becoming a projectile when the machine is turned on. If drilling flat material, position the table so that the hole in the center lines up with the bit, ensuring that the bit can pass through the table without cutting it. If drilling small or complex objects, place a table vise on the table and secure the workpiece into the vise's jaws. Align the material relative to the bit and secure the vise to the table with a clamp. Power on the machine and slowly lower the spindle until cutting is complete. Retract the spindle and power off the machine. When drilling metals, use cutting oil and frequently retract the spindle to clear chips from the bit. If necessary, power down the machine to clear the chips.

Drilling holes in the center of a round object is difficult (Figure 2-97). An easy method for tackling this problem is to secure the material in the chuck and then lower the spindle and secure the material into the table vise. Fix the table vise in place with a clamp, loosen the chuck, and then retract the spindle. The material should now be perfectly aligned with the spindle and ready for drilling.

Figure 2-96. *Always adjust the spindle speed appropriately for the material and secure your work to the drill press's table prior to drilling.*

Figure 2-97. *Center material in the vise by using the chuck as an alignment tool.*

Heat Sources

Makerspaces conduct a variety of tasks that require a focused heat source. These heat sources are capable of temperature ranges low enough to cause heat-shrink tubing to react, and high enough to start a fire. Because of their potential for causing unintended damage, it is recommended that these tools be used on a heat resistant surface and are unplugged when not in use.

There are two types of heat sources commonly found in the Makerspace, the heat gun and the hot plate. You can use heat guns for tasks that require intense focused heat, whereas hot plates are designed to heat vessels or objects through conduction.

The common heat source features are identified as follows:

Heating Element
   The heating element consists of a wound nichrome wire or other resistive element that generates heat when electric power is applied.

Thermostat
   The thermostat is an electromechanical switch that turns the heater on and off in an attempt to maintain a desired temperature. Some heat sources only have a selector switch that puts the heater into high or low power modes. These heat source types need to be monitored even more closely as they can reach temperatures well above the combustion point of many materials.

Heat Source Types
Heat Gun
   Heat guns use a fan that blows air across the heating element, resulting in a high temperature gust of air (Figure 2-98). Many heat guns feature a selector switch that sets the power output of the gun. The power consumption and temperature range are typically illustrated on the heat gun's label.

Figure 2-98. *Heat gun.*

Heat Gun Use
   Heat guns operate by sliding the selector switch to the desired heat range and directing the hot air at the workpiece (Figure 2-99). You can use this tool by hand or set it vertically on a heat resistant surface for more stable use and when cooling. Ensure that the device is unplugged from the power source after the task is complete to reduce the risk of incident. Do not lay the heat gun on its side when cooling because the tip of the gun might still be hot.

Figure 2-99. *Heat heat-shrink tubing with a heat gun to evenly shrink it.*

Hot Plate

Hot plates use a resistive element enclosed in a conductive metal plate (Figure 2-100). When power is applied to the element, the resulting heat is spread out over the extents of the conductive plate. Many hot plates feature an indicator that illuminates when power is applied to the element.

Hot Plate Use

You use hot plates by setting the desired temperature on the thermostat dial and applying power to the device (Figure 2-101). When the desired temperature is reached, the indicator lamp will turn off and on with respect to the heating of the element. Ensure that the power is removed from the device after the task is complete to reduce the risk of incident.

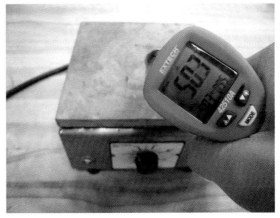

Figure 2-101. *Thermostats are often inaccurate, so use an infrared thermomenter to read the surface temperature.*

Figure 2-100. *Hot plate.*

Clamping, Joining, and Measurement

3

Although hand and power tools give us the means to construct and deconstruct, the tasks that they tackle are supported by a wide range of clamps, fasteners, adhesives, and measurement tools. Each category of support tool makes it possible for work to be conducted in a more secure and safe manner. They are mechanically and chemically bonded, and checked for accuracy.

Clamps and Vises

Clamps and vises provide a temporary mechanical bond between two or more materials and are widely used during the construction of a project. These tools feature robust and unique designs that can secure objects and materials at almost any angle, making the construction process eas-

ier to accomplish. Every Makerspace should offer a wide range of clamps and vises to ensure the proper construction of its ongoing projects.

Clamps

Clamps provide a compressive force to secure one or more objects in place while work proceeds on them. Regardless of design, clamps are meant to function only as a temporary support mechanism and should be removed prior to project operation.

The common clamp features are identified as follows:

Opening
> The opening refers to the maximum thickness of material that can be secured.

Reach
> The reach is the maximum depth at which material can pass through the clamping mechanism.

Clamp Types
Bar
> Bar clamps are designed to compress materials over a wide range of thicknesses (Figure 3-1). These clamps feature a sliding head that can be quickly positioned into place and compressed with a trigger. To release the clamp, simply press the release button and slide open the jaws.

Figure 3-1. *Bar clamp.*

Figure 3-3. *Spring clamp.*

C-clamps

C-clamps are the most utilitarian type of clamp due to their ease of use and clamping power (Figure 3-2). The clamp consists of a C-shaped body with a threaded hand screw. There is usually a swivel foot attached to the base of the screw that helps to prevent the clamp from "walking" while it's being tightened.

Figure 3-2. *C-clamp.*

Spring

Spring clamps are the quintessential temporary clamp (Figure 3-3). Similar in operation to binder-clip clamps, these devices secure material through compression produced by an internal spring. They are great for flattening and straightening thin material or for temporarily securing light materials in place.

General Clamp Use

You can use clamps by positioning one or more pieces of material in the opening and compressing the clamping mechanism. The amount of compressive force depends on the clamp type and application. When using a clamp on soft materials or those with a sensitive surface finish, place a wood block between the material and the clamp's feet (Figure 3-4). This way, the clamp can compress the material without marring the surface.

Figure 3-4. *Place a piece of wood between the clamp and your material to prevent damage.*

Vises

Vises operate by using a screw mechanism to compress and secure objects and materials in their jaws. You can use these tools to secure material that you can then work on by hand or machine, making them an important addition to any Makerspace.

Base

The base of the vise is the stationary component that is mechanically coupled to the work surface (such as a drill press table). Bases often include special mounting features that facilitate easy mounting. Avoid using vises that are mounted by using a C-clamp-style mechanism. These have the tendency to shift while working and can damage the work surface.

Jaw

A vise's jaws consist of two pieces: the first is attached directly to the base and does not move; the second is attached to the screw mechanism and you can compress and retract it by rotating the vise's handle. Some jaws include special cutouts that are designed to align and hold round objects. These features are helpful when attempting to drill a round object or hold it in place while cutting.

When installing a vise on a work surface, use bolts rather than screws. Pass the bolt through the base of the vise and secure it in place with a washer, lock washer, nut combination. This setup will help prevent the vise from shifting over time.

Vise Types

Bench

Bench vises are designed to mechanically couple to a work surface and provide a robust method for securing material on which you need to work (Figure 3-5). You should install a vise at the height of the average person's bent elbow. This positioning helps reduce fatigue when being used. Bench vises feature a large set of jaws that are compressed by turning the handle. When not in use, make sure the handle points down, reducing the risk of someone walking into the outstretched handle.

Figure 3-5. *Bench vise.*

Table

The table vise is designed to secure small and complex material in place while being drilled (Figure 3-6). This vice features a wide base and bolt mounts that can be used to both clamp and bolt the vise to the drill press table. Ensure that the vise is securely mounted to the table. This prevents the vice from becoming a projectile if the drill bit catches while drilling.

General Vise Use

Because vises are intended to compress and fix objects and materials in place, their use is relatively cut and dried (Figure 3-7). Open the jaws of the vise wide enough to accept the material and then compress until secure. Soft material can be easily damaged by the jaws of the vise if over compressed. To prevent this, cover the jaws with brass or wood.

Figure 3-6. *Table vise.*

Figure 3-7. *When cutting material, cut as close to the vise as possible to prevent vibration.*

Adhesives and Fasteners

The world is held together with adhesives and fasteners. Adhesives are compounds that produce both temporary and permanent bonds that secure, seal, and mask. Fasteners are mechanical devices that compress material together using a simple machine. Fasteners produce both temporary and permanent bonds by means of a mechanical advantage.

Adhesives

Adhesives are the chemical alternative to a fastener's mechanical bond. There are many types of adhesives designed for specific applications ranging from simple wire isolation to bonds meant for extreme temperatures.

Adhesive

The adhesive refers to the actual material component that is responsible for bonding.

Backing

The backing is the film on which the adhesive is applied, allowing for two or more materials to be bound together. Most backings are made out of plastic or natural materials and serve to provide tension support to the adhesives bonding power.

Tapes

It is important to note the environmental conditions your tape will encounter in order to ensure its optimal performance. Use Table 3-1 to help you decide which type of tape will work best for your project.

Table 3-1. The right tape for the job

Type	Outdoor	Wet	Dry	Hot	Cold
Cellophane	No	No	Yes	No	Yes
Duct	Yes	Yes	Yes	Yes	Yes
Electrical	Yes	Yes	Yes	No	Yes
Masking	No	No	Yes	No	Yes
Polyimide	Yes	Yes	Yes	Yes	Yes

Cellophane

This general tape type is designed for use in light-duty situations (Figure 3-8). It consists of a thin plastic or cellophane backing that has been coated with single-use, pressure-sensitive adhesive. Because this tape type is not meant to be removed, the tape is easily torn and often leaves behind adhesive residue.

Figure 3-8. *Cellophane tape.*

Figure 3-9. *Duct tape.*

Duct

Duct tape certainly has a reputation for being the "fix-all" adhesive for virtually any project (Figure 3-9). This tape comprises a fiber-reinforced polyethylene backing with a rubber-based adhesive and features excellent tack characteristics. New advances in duct tape technology have attempted to solve the material degradation problem experienced when duct tape is continuously exposed to the elements. This adhesive, once solely used for installing and mending ventilation systems, has become a humorously useful universal tool.

Steve Smith's "The Red Green Show" (http://www.redgreen.com) is a wonderfully comedic do-it-yourself sitcom where a majority of the projects involve the use or misuse of duct tape.

Electrical

Electrical tape is another type of pressure-sensitive adhesive that is electrically non-conductive, making it great for insulating exposed wires (Figure 3-10). This tape type consists of a PVC backing material coated with a waterproof rubber resin adhesive. Depending on the brand, electrical tape is designed to withstand environmental temperatures from −18°C to 105°C and will not degrade under exposure to ultraviolet (UV) light. All this being said, heat-shrink tubing should be used in lieu of electrical tape whenever possible because electrical tape tends to lose its holding power and become unstuck over time.

Figure 3-10. *Electrical tape.*

Masking

Masking tape is a great adhesive for projects that only require temporary joining (Figure 3-11). This tape is generally made from paper backing material that is coated with a gummy adhesive that conforms to surfaces and can be removed without leaving residue. These characteristics make masking tape an ideal adhesive for temporarily securing parts while gluing, masking off areas for painting, and labeling components. Blue painter's masking tape actually functions as a good build surface for ABS 3D printing.

Figure 3-12. *Polyimide tape.*

General Tape Use

The elements and working environment can stress tape resulting in it losing its grip. Also, tapes like duct and electrical deteriorate over time leaving a sticky residue behind (Figure 3-13). When using tape to secure your work, don't touch the adhesive as the oils on your skin can lessen its grip. Lay the tape flat across the surface and gently rub the backing to promote adhesion.

Figure 3-11. *Masking tape.*

Polyimide

Polyimide, or Kapton tape, has become increasingly popular in the Maker world due to its high temperature tolerance, electrical insulation, and excellent tack characteristics (Figure 3-12). This tape type consists of a polyimide material backing with a silicon adhesive. Polyimide tape also functions as an excellent build surface material for ABS and PLA 3D printing.

Figure 3-13. *Use the correct tape for the job.*

Glues

It is important to note the environmental conditions your glue will encounter and the surfaces it will be used on in order to ensure its optimal performance. Use Table 3-2 to help you decide which type of glue will work best for your project.

Table 3-2. The right glue for the job

Type	Wet	Dry	Hot	Cold	Wood	Plastic	Metal	Strength
Hot melt	Yes	Yes	No	Yes	Yes	Yes	Yes	Medium
Super glue	Yes	Yes	Yes	Yes	Yes	Yes	Yes	High
White	No	Yes	No	No	Yes	No	No	Low
Wood glue	No	Yes	No	No	Yes	No	No	Medium

Hot Melt

Hot melt is a solvent-free thermoplastic adhesive that works in conjunction with a heating element to liquefy the feedstock for application (Figure 3-14). The adhesive then quickly sets after cooling, producing a strong bond on a variety of materials including wood, fabric, and plastic.

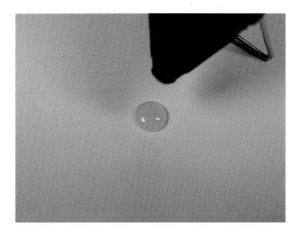

Figure 3-14. Hot melt adhesive.

Super

The name says it all (Figure 3-15). Super glue, or cyanoacrolate, is a quick-drying general purpose adhesive that is good for bonding metals, rubber, plastic, ceramics, and wood. This adhesive type is designed to cure within 20 to 30 seconds, depending on viscosity, and can be dramatically decreased with the addition of a spray-on accelerator. The cur-

ing of super glue is an exothermic reaction, which means it produces heat when curing. Ensure that you are only gluing materials that can withstand the heat generated without deformation or consider using a slower-cure super glue if you are unsure. Super glue is prone to quickly polymerize on skin contact; however, you can easily remove it with acetone or nail polish remover.

Figure 3-15. Super glue.

Do not use super glue anywhere you require transparency, such as on the lenses of a pair of glasses. The volatiles in super glue have a tendency to creep across smooth surfaces and can cloud them with a semi-permanent white film. So do not use super glue to fix your glasses!

White

White glue, or general purpose glue, is a staple product for any primary-school classroom (Figure 3-16). This inexpensive adhesive made from polyvinyl acetate is designed for use on porous materials that will not come in contact with water. Because this glue is slow-setting and exhibits relative low strength, it is only recommended in a pinch. After application, the project should be clamped for 30 minutes and allowed to cure for 24 hours.

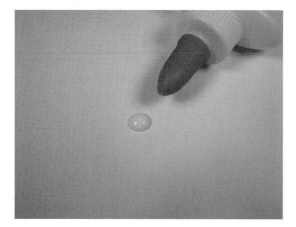

Figure 3-16. *White glue.*

Wood

Wood glue is an excellent alternative to white glue because it forms a stronger bond, dries quicker, and can be weatherproof (Figure 3-17). This glue type is made from an aliphatic resin with a slightly yellow appearance. After application, the project should be clamped for 30 minutes and allowed to cure for 24 hours.

Figure 3-17. *Wood glue.*

General Glue Use

Glue is designed for one purpose, to adhere two or more objects together by bonding to the material's surface. This is accomplished by first preparing the surface to better adhere to the glue. For porous surfaces like wood, not much preparation is needed because the glue has a ready-made means for gripping the material. Other materials, like metal and plastic, require surface preparation prior to gluing. First remove any oil or residue from the surface of the material with alcohol and wipe clean. Lightly scour the surface with medium-grade sandpaper in a criss-cross pattern. The scratches made by the sandpaper will give the glue a better surface to which to adhere. Apply the glue to each component and press together. Glue works best if left undisturbed and in compression (Figure 3-18) for the entirety of the cure time. Disturbing or not pressing the joint together will result in a poor glue joint that will ultimately fail.

Figure 3-18. *Glue works best when it is compressed and undisturbed until it sets.*

Specialty Glue

Use Table 3-3 to help you decide which type of specialty glue will work best for your project.

Epoxy

Epoxy is a two-part adhesive comprising resin and hardener (Figure 3-19). You measure the two liquids to a predetermined ratio and throughly mix before applying. This initiates an exothermic reaction that takes anywhere between 5 and 30 minutes to fully cure, depending on type. You can use epoxy to adhere a variety of materials, including wood, plastic, metal, and glass. Certain formulas of epoxy, like JB Weld, can withstand high temperatures and can even be used to repair engine components!

Figure 3-20. *Mix epoxy until the resin and hardener have fully combined.*

Epoxy produces an exothermic reaction when mixed, which can melt thin plastic cups.

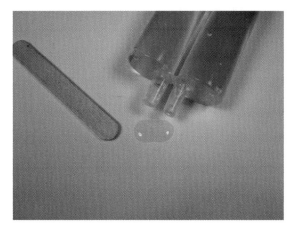

Figure 3-19. *Epoxy.*

Epoxy Use

Most epoxies follow a 50/50 resin-to-hardener ratio. To mix epoxy, begin by putting on a pair of gloves and measuring equal amounts of hardener and resin into two separate cups. If needed, you can make these measurements using a scale. Pour the hardener into the resin cup and mix for 1 minute. Ensure that both compounds are completely mixed (Figure 3-20) before using; poorly mixed epoxy will not properly set. Apply the mixed epoxy to your project, but don't throw away the mixing cup. You can use this cup as a representative sample to determine the state of the solidifying epoxy without disturbing the project.

Thread-Lock

Thread-locking adhesive is a vital component for securing nuts and bolts when used in situations that experience any amount of vibration or impact that might loosen the fastener (Figure 3-21). Thread-locking adhesive is an anaerobic compound that prevents this loosening by filling the small gaps found between a fastener's interlocking threads. The adhesive then bonds to the mating interface and produces either a permanent or semi-permanent bond. Thread-locking adhesive is typically color coded to indicate the strength of the bond, as listed in Table 3-4.

Table 3-3. The right specialty glue for the job

Type	Wet	Dry	Hot	Cold	Wood	Plastic	Metal	Strength
Epoxy	Yes	Yes	Yes	Yes	Yes	Yes	Yes	High
Thread-lock	Yes	Yes	Yes	Yes	No	No	Yes	High

Figure 3-21. *Thread-locking adhesive.*

Figure 3-22. *Always clean the threads of the fastener prior to applying thread-lock.*

Table 3-4. *Thread-locking adhesive color code*

Color	Locking Shear	Notes
Green	1500 PSI	Low viscosity allows for application without removing fastener
Blue	1200 to 1500 PSI	Low shear permits fastener removal
Red	2500 to 3500 PSI	Only removable by heating fastener to 260° C

Thread-Lock Use

Before using a thread-locking adhesive, you must clean the threads of the fastener to remove any compounds that might prevent adhesion (Figure 3-22). Place a small drop of thread-lock onto the threads of the fastener and secure in place with a washer and a nut. The adhesive will bond within the manufacturer's stated time and can only be removed by following the color-coded procedure.

Fasteners

Fasteners provide a mechanical coupling between two or more objects and feature a wide range of sizes and designs. The earliest fasteners were forged by hand and revolutionized construction projects, making bigger and stronger structures possible. Today, Makerspaces rely on a wide range of threaded fasteners in virtually every project.

The best rule of thumb for determining how to loosen or tighten a fastener is the age-old saying "righty-tighty lefty-loosey." This saying illustrates the angle of the fastener's threads allowing for the fastener to tighten when turned clockwise and loosen when turned counterclockwise.

The common fastener features are identified as follows:

Head Style

The threaded fastener's head style dictates both the type of tool used to drive the fastener as well as how it is secured. Each head style is designed to optimize the fastener's capability.

Thread Size

The thread size specifies both the diameter of the fastener and the quantity of threads per inch or millimeter. Standard-sized fasteners are identified using a number scale for sizes under 1/4 in and fractional for sizes above that. Metric fasteners are identified with a similar number scale, although the metric unit is proceeded by an uppercase *M*. Both standard and metric fasteners also include a thread-per-unit identifier that proceeds the diameter. A 4-40 standard screw will measure 0.112 inch in diameter and has 40 threads per inch.

Diameter

The most important feature of any fastener is its diameter, which dictates both the size of the hole it can fit as well as the type of threads it can support. Each fastener is designed to mate with either a hole or fastener that corresponds to its diameter. If a fastener is used whose diameter is too small for the hole or fastener, the fastener will loosen over time, produce a poor coupling, or fail entirely.

Rivets and Grommets

Blind Rivet

A blind rivet, or pop rivet, is a nonthreaded mechanical fastener that produces a permanent coupling between thin metal, wood, and plastic (Figure 3-23). When these fasteners are secured, their body is physically deformed as a nail-shaped mandrel is pulled into its body, producing a quick and strong joint. When the rivet is fully compressed (Figure 3-24), the mandrel shears off, leaving only the rivet behind.

Blind rivets can fasten a variety of materials depending on their composition. Table 3-5 illustrates different rivet material types, their diameter, and the type of material they can secure.

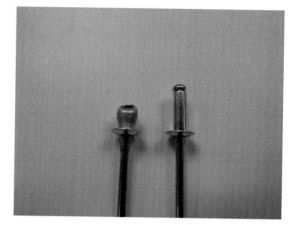

Figure 3-23. *Blind rivets.*

Figure 3-24. *After the rivet is compressed, a small piece of the mandrel is left behind, securing the fastener.*

Rivet Gun

Rivet guns are used to secure blind rivets in place by grabbing and pulling the mandrel through the rivet after it has been seated (Figure 3-25). These tools are accompanied by a series of nosepieces that correspond to the rivet's mandrel diameter.

Table 3-5. Blind rivet compatibility

Rivet	Use
Aluminum	Aluminum
Copper	Stainless steel/copper/brass
Steel	Steel
Nylon	Plastic/wood/metal/fiberglass

Figure 3-25. *Rivet gun.*

Figure 3-26. *Firmly press the rivet against the material when compressed to ensure a tight joint.*

Rivet Gun Use

Begin by drilling an appropriately sized hole for the rivet in your material. If you are overlapping two or more pieces of material, clamp them together and drill them all at once. Attach the proper nosepiece to the gun and slide in a rivet. Push the rivet into the hole and squeeze the gun's handle (Figure 3-26). The mandrel should pull through the rivet and pop off. If it doesn't, simply squeeze the handle again until it does.

To remove a secured rivet, simply use a drill and bit that is slightly larger than the rivet's head and drill until the head pops off. Then, use a center punch to remove the remaining material. Be careful not to drill into your work!

Screws

Hex-Head Cap

Cap screws feature a hex-shaped head and are most commonly found in larger sizes (Figure 3-27). The robustness of the head and drive type makes it possible for cap screws to withstand a large amount of torque and resist wear.

Figure 3-27. *Hex-head cap screws.*

Machine

Machine screws feature one of many drive system types and are commonly found in smaller sizes (Figure 3-28). The recessed drive, rounded head, and high quantity of threads per unit make machine screws ideal for most projects.

Figure 3-28. *Machine screw.*

Set

Set screws are designed specifically for use in recessed areas (Figure 3-29). Because these screws do not have a head, they are only limited by the depth of the threaded hole. Most set screws are used to secure gears and pulleys onto shafts by driving into the shaft, securing the component in place.

Figure 3-29. *Set screw.*

Sheet Metal

Sheet metal screws feature an aggressive thread pattern that is designed to deform and bind to thin metal (Figure 3-30). These screws also work well with many types of wood and plastic.

Figure 3-30. *Sheet metal screw.*

Socket-Head Cap

Socket head cap screws feature a hexagonal drive enclosed in a round head (Figure 3-31). This design makes it possible for the screw to be entirely countersunk into the work-piece, providing for a more finished appearance.

Figure 3-31. *Socket-head cap screw.*

General Screw Use

Screws are used in two configurations: threaded into a hole and secured with a nut (Figure 3-32). To thread a screw into a hole, align the proper-sized screw at the opening so that it is parallel to the hole's wall. Counter-rotate the screw while applying a small amount of pressure until you feel a small click. This indicates that the threads of the screw have aligned with the hole's threads and you can begin tightening down the screw. Trying to drive a screw without first aligning the threads can result in cross-threading which occurs when the threads cross over each other, causing permanent damage. To secure a screw with a nut, slide the screw through the workpiece and place a washer over the exposed threads at the end of the screw. Attach a nut over the end and tighten until secure.

Figure 3-33. *Acorn nut.*

Captive

Captive nuts are designed to be embedded into material and function as an alternative to a threaded hole (Figure 3-34). These nuts feature a press-in mounting system that allows for the nut to mechanically interface with the desired material when compressed.

Figure 3-32. *Use a washer between the head and the nut to help prevent the fastener from loosening due to flexing and vibration.*

Nuts
Acorn

Acorn nuts feature a hex shaped drive and a domed covering over one side (Figure 3-33). This dome is designed to prevent accidental contact with the exposed threads; their primary purpose is safety, but they are also a decorative feature.

Figure 3-34. *Captive nuts.*

Hex

Hex nuts are the most commonly used nut and feature the same hex shape as cap screws (Figure 3-35). For situations in which the nut is prone to loosening, a second nut can be added to the fastener to "jam" the first in place, preventing movement.

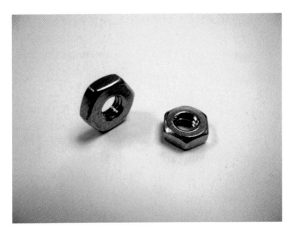

Figure 3-35. *Hex nuts.*

Figure 3-37. *Wing nuts.*

Lock Nut

Lock nuts are identical to hex nuts in shape, but include an embedded nylon locking ring at the top of the nut (Figure 3-36). When the nut is fully threaded onto a fastener, the nylon insert binds with the fasteners threads, preventing the nut from loosening.

Figure 3-36. *Lock nuts.*

Wing

Wing nuts feature two wing-shaped extensions with which you can twist the fastener on and off with just your fingers (Figure 3-37). Because wing nuts lack the standard hex shape, they are only intended for use on interfaces that do not require a large amount of compression.

Washers
Flat

Flat washers are simply stamped pieces of sheet metal that act as a barrier between the nut and the fastened material (Figure 3-38). Without a washer, nuts will deform the workpiece, making them more prone to loosening over time.

Figure 3-38. *Flat washers.*

Lock

Lock washers are used for situations in which the fastener is prone to loosening (Figure 3-39). There are many types of lock washers, including split ring, internal tooth, and external tooth, and all are designed to embed themselves in both the fastened material and the nut.

Figure 3-39. *Lock washers.*

Measurement

Measurement devices are designed to verify the dimensions of an object or to establish the dimensions of something that is to be created. By properly using dimensional measurement tools, you can verify and adjust your Makerspace projects.

Dimensional

You use dimensional measurement tools to perform quick and relatively accurate measurements. You can use these tools for both angle and length; thus they serve as the backbone of most design.

The common dimensional tool features are identified as follows:

Graduations

Dimensional measurement tools are divided into a series of graduations relative to that tool's measurement system. Standard, or English, measurement tools are typically delimited in inches, which are then subdivided to 1/16 in. Metric measurement tools are typically divided into centimeters, which are then subdivided down to the millimeter. Although there are dimensional measurement

tools that can measure to higher resolutions, they are typically done with tools design specifically for high-precision measurement.

Angular

Level

A level relies on the Earth's constant gravitational pull as its measurement reference (Figure 3-40). Each level consists of a straightedge (an unmarked rule) and one or more transparent vials containing a mysterious fluorescent-yellow fluid and a bubble. The vials are marked with two lines indicating the center of the vial as well as the width of the bubble. When the bubble is positioned in the center of the vial, the level is, well…level.

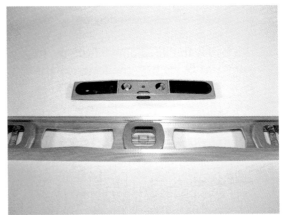

Figure 3-40. *Levels.*

Level Use

Levels are used by placing the tool onto the surface that is to be measured (Figure 3-41). Ensure that the level does not wobble and is securely positioned. Each level contains one or more bubble vials that are used to determine how level an object is. Looking at the vial, adjust the object until the bubble is centered between the two lines.

Figure 3-41. *A level indicates a level surface when the bubble is positioned between the vial's indicator lines.*

Figure 3-43. *Some protractors offer a guide that assists with measurement.*

Protractor

Protractors offer a means for measuring angles (Figure 3-42). This device contains a series of angle graduations, typically illustrating angle to the degree. Some protractors also have an attached rule that is used as a pointer to indicate angle while offering another measurement surface.

Linear

Rule

Rules are bars of wood, plastic, or metal marked with graduations in the unit of measure for the device (Figure 3-44). There are many types of rules that are designed to assist as much as possible in the measurement and identification of a material's width, length, and height. Precision or machinist's rules offer higher-resolution graduations and deliver much higher measurement precision.

Figure 3-42. *Protractor.*

Figure 3-44. *Rules.*

Protractor Use

To use a protractor, align the base of the protractor to the reference side and position the arm to the desired angle (Figure 3-43). This tool can be used to both measure the angle of an existing feature or provide a reference angle for marking a new feature.

Square

Squares, or *framing squares*, provide an accurate 90-degree measurement reference (Figure 3-45). Each side of the framing square displays graduations that extend from the root of the angle to each end. This tool can be used to provide perpendicular measurements and markings relative to a side as well as to verify that an angle is square.

Figure 3-46. *Tape rules.*

T-Square

T-Squares offer a head that is perpendicular (90 degrees) to the leg—the part of the t-square with marked graduations (Figure 3-47). This head serves as a reference for measuring and marking with higher precision. Although most heads do not contain graduations, variations like the framing square offer graduations on both legs.

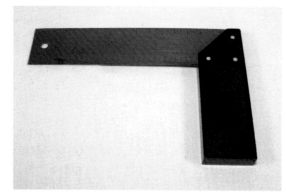

Figure 3-45. *Square.*

Tape Rule

Tape rules offer the same measurement method as a standard straight rule, but allow for much larger measurements (Figure 3-46). This device uses a polyester-coated spring steel blade that uncoils as the tape is extended. When the measurement is complete, a release is pressed and the tape rewinds back into its enclosure. Although tape measures are good at making large measurements, there is a lot of room for error in the measurement due to deflection of the tape and off-angle positioning. If a higher-precision measurement is required, a long solid rule is recommended.

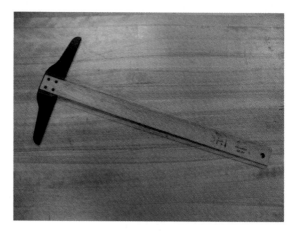

Figure 3-47. *T-square.*

General Linear Measurement

Each linear measurement tool features a series of graduations that indicate the distance between two points (Figure 3-48). Metric rules are typically graduated down to the single millimeter, while standard or English rules measure down to 1/16th of an inch. Some rules, like machinist rules, can measure distances as small as 1/2 mm or 1/64 in.

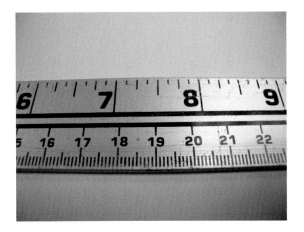

Figure 3-48. *Some rulers feature both metric and standard measurement graduations.*

Precision Linear
Caliper

A caliper can accurately measure the internal or outer diameter, thickness, or depth of an object (Figure 3-49). The caliper consists of six primary parts (Figure 3-50): the inside jaw, the outside jaw, the dial and lock screw, the fine adjustment, and the main slide.

Figure 3-49. *Dial caliper.*

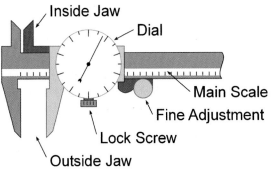

Figure 3-50. *Caliper components.*

Caliper Measurement

To make a measurement with a caliper, place the desired material between one of the measurement contacts, then use the fine-adjust roll to gently snug the contacts against the object. Next, read the last visible number on the main scale. Each large number on the bar represents 1 in and each small number represents 0.1 in. Read the indicator on the dial. Each dash represents 0.001 in. Combine the two numbers to get the measurement.

To calibrate the caliper, carefully slide the contacts together so that there is no gap (Figure 3-51). Next, carefully rotate the outside of the dial indicator until the "0" mark lines up with the indicator.

Figure 3-51. *Always calibrate (zero out) a caliper prior to use and use the appropriate measurement feature.*

Micrometer

A micrometer can accurately measure the thickness or width of an object with resolutions up to 1/10,000th of an inch or 1/100th of a millimeter (Figure 3-52). Micrometers contain a series of components that work together to produce very accurate measurement. The micrometer consists of three primary components: the anvil, spindle, and thimble (Figure 3-53).

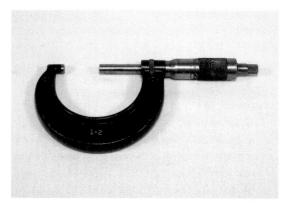

Figure 3-52. *Micrometer.*

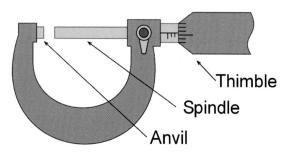

Figure 3-53. *Micrometer components.*

Micrometer Use

To take a measurement with a micrometer, place the desired material between the anvil and spindle and then carefully rotate the ratchet on the back of the thimble until the ratchet slips (Figure 3-54). Next, read the highest visible number on the sleeve. This number acts as the tenths decimal in your reading. Then, count the number of lines on the sleeve past the highest visible number.

Each line past the highest number counts as 0.025 in. Finally, read the number on the thimble that lies closest to the center line on the sleeve. Each mark indicates 0.001 in. Determine the measurement by combining all three numbers.

Figure 3-54. *Use the ratchet to ensure proper pressure and accurate measurement.*

Inspection

Inspection tools are used to investigate the operation of an object without interfering with its functionality. Both multimeters (Figure 3-55) and temperature measurement devices are integral to the Makerspace. Multimeters provide several methods for making electrical measurements. Although there are almost unlimited brands and types, they all include a standard set of tools. Most temperature measurements are made by using devices such as thermocouples, resistance temperature detectors (RTDs), thermistors, and solid-state sensors. Each of these devices have unique characteristics that can affect their accuracy, range, and use.

Figure 3-55. *Multimeter.*

Multimeter
AC/DC Voltage

Multimeters can measure a wide range of both AC and DC voltages. The range is selected using the dial or is done automatically. Ensure that the voltage you are measuring does not exceed the limits of your device. If you do, there is a fuse that will blow when the limits are exceeded. Typically, you can replace fuses easily by removing the back of the meter and dropping in a new one. Remember, voltage is measured in parallel with the supply.

Capacitance

The capacitance function provides value measurement and verification. You can also use this function to identify the polarity of unmarked polar capacitors. Some multimeters have separate sockets designed for capacitor measurement; others use the test leads.

Continuity

You use the continuity function to determine whether a circuit is open or closed. This tool can be quite useful for isolating a circuit fault or for reverse engineering. The continuity test also shows the resistance between the two leads and can help determine line losses and short circuits.

Current

Multimeters typically have two levels of current sensing 0–400 mA and 401 mA to 10 A. This is done by measuring the voltage drop over a shunt resistor that's a part of your multimeter. Using the 401 mA to 10 A function requires a reconfiguration of the test leads so that the red lead is positioned in the 10 A socket and the black lead is positioned in the ground socket. It is common to burn out the fuse on the current circuit when the test leads are incorrectly configured. This fuse can be replaced in the same manner as the multifunction fuse mentioned in the voltage sensing section. Remember, current is measured in series with the supply.

Diode

You can measure the polarity and voltage drop of a diode with this function. Connect the test leads to the diode's anode and cathode and the display will indicate the voltage drop. You can also use this to determine the functionality of the diode as well as its polarity.

Resistance

Resistance measurement extends beyond the standard measurement of resistor values. You can also use it for circuit analysis and to help in determining faults. This function operates by passing a small voltage between the two leads and measuring the drop. By using $V = IR$, you can deduce the resistance.

Frequency

Frequency measurement is a great tool to have because you can use it to determine the function of oscillators, signals, and other low-frequency sources. Depending on your multimeter, frequency analysis usually ranges from 0–10 MHz.

Inductance

Multimeters measure inductance by connecting a precision resistor in series with the inductor and measuring the junction voltage when a signal is passed through the circuit. The multimeter can then derive the inductance based on the response to the signal.

Temperature

Some multimeters ship with thermocouples and can measure a wide range of temperatures. Because the thermocouple is a polar device, make sure that the sensor is connected correctly.

Transistor

Multimeters can measure transistor hFE or forward current gain of NPN and PNP transistors (not MOSFETs). Depending on the multimeter, there might be a series of holes configured in a row and in a circle pattern that accommodates connecting to the transistor's base, collector, and emitter pins. Alternatively, the function of a transistor can be determined by using the multimeter's diode test function and connecting the test leads to the base/emitter and base/collector respectively.

General Multimeter Use

To use a multimeter, you first select the desired measurement tool by rotating the selection dial to the desired function and inserting the test leads into the appropriate sockets. These sockets are marked according to their function and typically consist of a high current, ground, and multifunction position. Depending on the design of your multimeter, many of the functions have range limits. Ensure that you adjust the dial (Figure 3-56) to the approximate range of your source. If your measurement exceeds the limits of that function, you will get the infamous "OL" message, indicating the inability to make an accurate measurement.

Figure 3-56. *Use the multimeter's dial to select the desired tool and measurement range.*

Temperature
Infrared

Infrared thermometers utilize a small infrared sensor that measures the surface temperature of an object without making contact. These devices are quite useful at measuring temperatures that exceed the limits of thermistors or are sensitive to physical contact. Because these sensors are susceptible to reflected infrared light, it is often difficult to measure the surface temperature of reflective objects such as glass and polished metal.

RTD

Rather than produce voltage with respect to temperature, the RTD changes resistance. This change in resistance can then be correlated to temperature, depending on the RTD's material type. Because platinum is commonly used due to its accuracy, RTDs often cost more but produce better results compared to its counterparts. The RTD element is available as either a glass vial, stainless steel probe, or surface element with a temperature range of –200°C to 850°C, depending on type.

Thermistor

A thermistor functions in a similar manner to an RTD but with less accuracy and temperature range. Thermistors are typically small, round elements with axial leads or a metal probe. They and feature a measurement range of 0°C to 100°C, depending on type.

Thermocouple

When two dissimilar metals or metal alloys are joined, they often generate a small voltage at the junction. This voltage can then be correlated to a temperature through mathematic deduction. There are four common thermocouple types that illustrate the means of calibration: J, K, T, and E. The K-type thermocouple is the most commonly used type and features a temperature range of −200°C to 1,250°C, depending on wire diameter. Table 3-6 illustrates common thermocouple temperature ranges:

Table 3-6. Thermocouple types

Type	Range	Error Limit
J	0°C to 750°C	2.2°C
K	−200°C to 1250°C	2.2°C
E	−200°C to 900°C	1.7°C
T	−250°C to 350°C	1.0°C

General Temperature Measurement

Temperature sensors need to be physically connected to the object to make an accurate measurement. This can be achieved by mechanically mounting the sensor onto the object using a fastener or adhering it with thermally conductive adhesive. Using tape or other nonpermanent bonding methods will result in inaccurate measurement because the sensor can shift, distorting the reading.

The Electronics Workbench | 4.

Even though you can easily transport soldering equipment, it is best to have a space designated for soldering. By setting up specific areas for soldering, project productivity and success will increase. The station should contain all of the equipment and materials necessary to successfully complete a soldering project. This includes an electrostatic protective (ESD) mat and strap, soldering iron or irons, soldering iron stands, multiple solder types and diameters, "helping hands" or a circuit board vise, desoldering and rework tools, adjustable lighting and magnifier, and a system for ventilating flux fumes. Additional equipment might be made available at the stations depending on the ethos of the Makerspace, including tools such as oscilloscopes, multimeters, and signal analyzers.

Whether you are working with pre-made circuits or designing your own, chances are you are going to need to solder. This metalworking process is most commonly used to bond electronic components to a circuit board, wire-to-board, wire-to-wire, and board-to-board. The skill required to achieve a proper solder joint has become the backbone of tinkerers and Makers alike and often serves as the starting point for getting new minds interested in Making. The act of following a schematic and soldering required components to a board exposes the Maker to the circuit's purpose and function, eliminating the "black-box" mystery. There might still be mystery as to why the components work the way they do, but with anything technical, the more one works with components, the more they are able to identify the different component types and the packages in which they are enclosed. The act of soldering a circuit and the equipment involved has become integral to any Makerspace. By understanding the variety of soldering irons, solder types, and soldering methods the Makerspace will be more able to successfully tackle any soldering project.

Connect your soldering iron to an outlet timer. This ensures that they are not left on at the end of the day.

Soldering is the process in which two similar or dissimilar base metals are joined together by using metal fillers that have a relatively low melting point (typically 183°C to 361°F). The metal filler, or solder, is made up of metal alloys that exhibit properties necessary for maintaining stability and electrical conductivity in a wide array of environments. The process of soldering differs from welding in two fundamental ways: first, the two base metals never reach their melting points; and second, you can bond dissimilar metals, which is difficult or impossible to do with welding. This chapter discusses many of the hardware types and soldering methods with which you can solder virtually any component type.

Soldering Irons and Tip Types

A soldering iron is the baton Makers use to compose their masterpiece projects. And as any musician knows, there is never just one instrument in an orchestra. The professional solder technician relies on a vast array of soldering irons and tip types, each designed for very specific tasks. They have been custom tailored to assist the technician in repeatedly producing the best possible solder joint in the least amount of time. For the hobbyist, most through-hole soldering can be completed with a simple iron and your favorite tip, while soldering surface-mount compo-

nents requires slightly more sophisticated equipment.

Look for soldering irons that are "ESD Safe." This ensures that the soldering iron tip will not allow for electrostatic energy to discharge into your circuit.

Soldering Iron Types

Soldering irons are available to satisfy many different types of soldering requirements, offering a wide array of features and capabilities. In general, irons can be classified as fixed-temperature or adjustable. Fixed-temperature soldering irons often cost less than their adjustable counterparts because they require fewer components to operate. These irons work simply by passing electrical current through a heating element that is mechanically connected to the iron's tip. Adjustable soldering irons utilize a temperature feedback circuit and a controller to accurately control the tip's temperature.

Fixed-Temperature Type
Fixed-Temperature Soldering Iron

The fixed-temperature soldering iron or "fire starter," as it's also known, does not have a method for monitoring tip temperature and is only recommended for light, through-hole soldering (Figure 4-1). Heat is brought to the tip via a mechanically coupled heating element (Figure 4-2).

Fixed-temperature soldering irons are sold based on wattage, which loosely correlates to tip temperature. Some irons feature a wattage selector tip that affords better thermal control. Table 4-1 is a wattage to temperature comparison that helps to illustrate the appropriate iron for the task.

Figure 4-1. *Fixed-temperature soldering iron.*

Figure 4-2. *Fixed-temperature soldering iron cross section.*

Table 4-1. *Approximate watt-to-temperature comparison*

Wattage	Unloaded Temperature (°C)
15	275
25	350
40	400

Soldering Gun

When soldering components with a large thermal mass, like a large screw terminal or board-mount heat sink, it might be necessary to rely on a "heavy-duty" fixed-temperature soldering iron (Figure 4-3). These irons, also known as soldering guns, apply a large amount of power (200 W) to the tip to bring objects with a large thermal mass up to soldering temperature. Because these irons can reach temperatures in excess of 500°C, it is imperative that the joints be soldered as quickly as possible. Temperature this high will result in irreparable damage to the circuit board and adjoining components.

Figure 4-3. *High-power soldering gun.*

Adjustable Type
Actively Adjustable

Adjustable soldering irons use relatively simple technology to control the temperature at the tip and result in a much more consistent and successful soldering experience (Figure 4-4).

Though this type of iron can be more expensive than the fixed-temperature type, the benefits more than make up the difference. Adjustable irons operate by positioning a temperature sensor near the heating element (Figure 4-5). This sensor then sends information to a controller that cycles power to the heater, regulating temperature. More sophisticated controllers utilize a tip selection feature, which provides more accurate regulation by applying a temperature compensation offset, depending on the tip type.

After the desired temperature is selected (Figure 4-6), the controller then sends power to the iron's heater until the temperature is achieved. This type of iron tends to heat up quicker than fixed-type irons because it utilizes a higher-wattage heater.

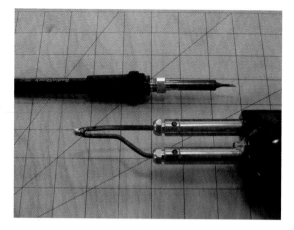

Figure 4-4. *Adjustable soldering irons.*

Figure 4-5. *Adjustable soldering iron cross sections.*

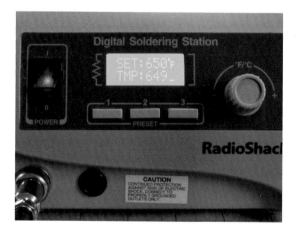

Figure 4-6. *The tip temperature is set by dialing in the desired temperature.*

Induction

An alternative design to the actively adjustable soldering iron uses radio waves to hold temperature (Figure 4-7). This process relies on induction to elevate the tip temperature to the material's *Curie point* and provides very stable thermal control. Since it is heated through induction, it takes only a matter of seconds for the tip to reach operational tem-

perature. As the tip actively solders, rather than cool down, the RF input automatically increases the power to the tip and results in a very consistent and efficient soldering experience.

Figure 4-7. *Induction-based soldering irons.*

Soldering Iron Tips

Every soldering iron transfers heat to the solder joint through a thermally conductive tip. This tip acts as a heat pipe, bringing the heat from the iron's heating element to the end of the tip. Tips (Figure 4-8) are available in a wide array of styles that are designed to meet almost any soldering or desoldering task.

Figure 4-8. *A cross section of a typical soldering iron tip.*

Most tips contain a thermally conductive copper core that is surrounded by a layer of iron. The iron provides the tip with rigidity and the desired solder bonding characteristics. Because iron is susceptible to corrosion, the outer surface of the tip is plated with chrome or nickel and only a small amount of iron is left exposed at the end.

Soldering iron tips can last for years when they are properly maintained. While in operation, the iron jacket is prone to deterioration due to accelerated rusting from the high heat. Adding a small amount of solder to the tip of the iron helps prevent this oxidation and ensures good thermal transfer to the joint. This process, known as *tinning* (Figure 4-9), also helps to prevent the natural corrosion that occurs when the iron is not in use. To maintain optimal performance, the tip should be cleaned frequently by wiping it on a wet sponge or brass pad (Figure 4-10). This process prevents the buildup of solder and burnt flux, which inhibit good joint formation.

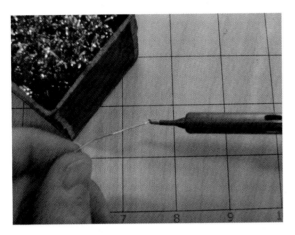

Figure 4-9. Soldering iron tips work better and last longer when tinned using a small amount of solder.

Figure 4-10. Some common tip cleaning tools.

Tip Types

It is important to choose the right tip for the job. Their size and shape dictate how much heat can be carried to the end of the tip. Larger tips are designed to heat up components with a large thermal mass, whereas smaller tips are the opposite. Soldering iron tips can be broken into three types: bevel, chisel, and conical.

Bevel

The bevel-type soldering iron tip serves as a good multipurpose solution (Figure 4-11). This tip features a large body diameter that tapers to a small beveled tip, which gives you the ability to perform satisfactory soldering of through-hole as well as some surface-mount device (SMD) components. You can rotate the tip to the side opposite the bevel to solder through-hole, and the tip of the bevel is optimal for soldering SMDs.

Figure 4-11. Bevel-type tip.

Chisel

The chisel-type, or screwdriver, tip is most commonly supplied with new soldering irons (Figure 4-12). This tip type is designed to provide a contact area perpendicular to the component, making it ideal for soldering through-hole and larger components.

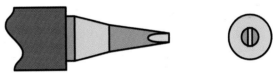

Figure 4-12. Chisel-type tip.

Conical

The conical-type tip is designed to focus the iron's heat to a very fine point and is optimized for soldering small through-hole and SMD components (Figure 4-13). These tips are prone to deterioration due to their fine point and should be cleaned and tinned regularly.

Figure 4-13. *Conical-type tip.*

Project: Solder Holder and Iron Stand

Most soldering irons come with flimsy stands that eventually break and offer little stability. With a quick trip to the plumbing aisle of your local home improvement store, a robust stand can be easily constructed from pieces of iron pipe.

This stand is designed to support a heated soldering iron. Make sure that it is securely coupled to a work surface to ensure that it does not fall over during use.

Materials

Materials List		
Item	**Quantity**	**Source**
1/2 in × 2.5 in iron pipe	1	Home improvement store
1/2 in × 4 in iron pipe	1	Home improvement store
1/2 in iron pipe cap	1	Home improvement store
1/2 in iron pipe flange	1	Home improvement store
1/2 in iron pipe "T"	2	Home improvement store

Procedure

Step 1

Assemble the stand as shown (Figure 4-14). Upon completion, use fasteners to mount the stand to the work surface, or to a 12 in × 12 in piece of plywood for a portable installation.

Figure 4-14. *The completed soldering iron stand.*

Step 2

Congratulations! You have just assembled a robust solder holder and iron stand. With something this simple, there's no reason every soldering station in your Makerspace should be without one.

Replace the T-junction with a 4-way adapter and screw in a 12 in piece of pipe. This extension can act as a great mount for a work light or magnifying glass. Before attempting this upgrade, ensure that the stand is mechanically coupled to the work surface.

Solder and Flux

There are an amazing amount of solder types, thicknesses, and fluxes available for soldering and desoldering electronic components. Solder is manufactured as either a bar, wire, or paste and can be found with or without flux. Until recently, most solder contained an alloy of a tin/lead or silver/tin/lead because they featured a low melting point and resistance to corrosion (Table 4-2). Now, most modern electronics use a lead-free alloy due to its lower environmental and material handling hazards.

Table 4-2. Common solder alloys

Process	Alloy	Melting Temperature (°C)	Form	Notes
Leaded	Sn60Pb40	183–190	Wire & bar	Provides a wide working temperature range and is easy to rework
Leaded	Sn63Pb37	183	Wire, bar & paste	Most common alloy and exhibits virtually no plastic range, also known as being eutectic
Lead free	Sn96.5Ag3.0Cu0.5	217–220	Wire, bar & paste	Most common lead-free alloy, provides rapid melting and high joint strength

When solder is heated and deposited, oxides form on the solder's surface and inhibit joint formation. This oxidation can lead to erratic circuit behavior or even complete joint failure. The inclusion of solder flux into the process helps to remove impurities from the connection and promotes proper joint wetting.

Do not use soldering flux that is designed for plumbing purposes. It can contain harmful corrosives that will physically damage your circuit board as well as being slightly electrically conductive.

Solder flux is available in three main formulas: rosin-based, no-clean, and water soluble. Each formula offers a wide range of viscosities, from a paste, which is good for tinning wires, to a near water-like consistency. The choice of solder flux reflects personal preference, and making multiple formulas available in your Makerspace is certainly a good way to make everyone happy.

Solder

Paste

Solder paste is extensively used for soldering SMD components and comes as a thick grey paste containing tiny solder particles suspended in a volatile flux medium (Figure 4-15). In a process known as reflow soldering, solder paste is applied to the surface of a bare circuit board using stencil and squeegee. The paste then acts as a mild adhesive, holding the components in place prior to the soldering process. When the board, paste, and components are heated, the solder melts and the flux evaporates leaving a precisely controlled solder joint. The advent of solder paste has made possible more complex soldering methods including double-sided circuit boards and pads underneath components (Table 4-3).

Figure 4-15. *Solder paste.*

Table 4-3. Recommended solder paste

Process	Flux Type	Alloy	Notes
Leaded	No-clean	Sn63Pb37	Good general purpose paste
Lead free	No-clean	SN96.5Ag3Cu0.5	Slightly more expensive then leaded and requires reflows at a higher temperature

Wire

Solder wire is the most common form of solder and is mainly used when soldering components by hand (Figure 4-16). The wire comes in two forms: solid and flux core, which indicates whether the solder contains a core of rosin/flux. Typically, rosin-core solder (Figure 4-17) is preferred as it eliminates the need for manually adding flux while soldering. Solid-core solder is typically used in situations that are sensitive to flux vapors or as per personal preference of the user.

Solder wire comes in a wide range of diameters (Table 4-4), ranging from 0.015 in to 0.125 in and is sold by weight. It is nice to have a range of solder diameters on hand because each lends itself to a different type of soldering.

Figure 4-16. Solder wire.

Figure 4-17. Solder spool specification label.

Table 4-4. Recommended solder wire diameters

Diameter (mm/in)	Application
0.39/0.015	Used for soldering surface mount components and signal wire
0.6/0.025	General purpose solder for through-hole components and light connectors
0.79/0.031	Used for large contacts and tinning/joining wires

Flux
Rosin

Rosin-based fluxes come in three flavors: nonactivated (R), mildly activated (RMA) and activated (RA). Nonactivated flux, whose production happens to be the project for this section, does not contain any additional acids that would require cleanup after soldering. RMA and RA fluxes contain acid activators that must be removed to prevent damage due to their corrosive nature. Regardless of the formula, rosin-based fluxes leave behind deposits that are difficult to remove. Commercial methods for removal require soaking the completed circuit in a bath of potentially harmful solvents that dissolve away the flux residue. Next, the boards are cleaned with deionized water and then baked to remove remaining moisture. The flux you'll make in this section's project can simply be left on the board or cleaned up with isopropyl alcohol and a toothbrush.

No-Clean

No-clean flux is a mildly activated flux that does not require removal after soldering. Once the flux dries, it leaves a noncorrosive and nonconductive residue behind that can easily be removed with isopropyl alcohol. As a result, no-clean flux is thin and evaporates quickly, especially if the board is warm.

Water Soluble

Water soluble flux is highly activated and can be removed with water. Due to its high acid content, water soluble flux must be cleaned from the entirety of the circuit board before put into operation. This cleaning process involves submerging the circuit board in a bath of deionized water and throughly scrubbing it to remove any residue. A failure to remove 100 percent of the flux can result in erratic operation or even permanent damage to the circuit.

Project: Two-Ingredient Soldering Flux

You really can never have enough flux! As a Makerspace increases in popularity, the need for cost-effective consumables becomes ever more apparent. It just so happens that you can make a lot of flux with just two low-cost ingredients.

One of the nice things about this project is that you can custom tailor the viscosity of your flux by adjusting the rosin-to-flux ratio. Table 4-5 il-

lustrates ratios that I have found to work well for the given situation.

Table 4-5. Rosin to 91 percent isopropyl alcohol ratio

Rosin (g)	Alcohol (mL)	Result
1	1.9	Thick
1	2.4	Standard
1	3.0	Thin

This project requires safety glasses that should be worn throughout its entirety.

Follow the safe handling procedures for isopropyl alcohol and use in a well ventilated area.

Materials

Item	Quantity	Source
Air-tight glass jar	1	Grocery store
>90% isopropyl alcohol	1	Grocery store
Natural rosin crystals or equivalent	1	Musical instrument supplier
Coffee filter	1	Grocery store
Scale	1	Grocery store
Graduated cylinder or equivalent	1	Scientific supply store

If natural rosin is not available, musicians' rosin makes a good alternative. You can find it at any local music store.

Makerspace Tools and Equipment

Hammer

Safety Glasses

Procedure
Step 1

Start by weighing out the predetermined amount of rosin onto a piece of paper

(Figure 4-18). Take the rosin out of the packaging and wrap it in a piece of paper. Put the bundle in a sturdy plastic bag and pulverize with a hammer. The finer you pulverize the rosin, the quicker it will absorb in the alcohol.

Figure 4-19. *Adjust the alcohol-to-rosin ratio until the desired viscosity is achieved.*

Figure 4-18. *Measuring out your rosin.*

Step 2

Determine your desired viscosity, measure out the appropriate amount of alcohol as shown in Table 4-5, and then add it to the jar. Carefully pour the crushed rosin into the jar and seal the lid. Ensure that the lid is on tight because the alcohol can evaporate quickly. Gently swirl the mixture until the rosin is fully dissolved. Let the mixture sit for a few hours or overnight until complete.

Step 3

When the rosin has fully dissolved into the mixture, place the funnel and coffee filter into the second jar. Carefully pour the initial mixture through the filter and into the second jar. This process eliminates any large particles that might hinder your soldering (Figure 4-19). After soldering, clean up flux residue with isopropyl alcohol, flush the board with distilled water, and then dry it using compressed air.

Through-Hole Soldering

Through-hole soldering is the most common type because it's not difficult to learn and doesn't require exotic soldering equipment. Through-hole describes the mounting configuration of an electronic device containing one or more leads. These leads are configured such that a solder connection can be made on the side of the circuit board opposite that of the component. Their relatively large size, easy handling, and ability to be used with a breadboard has made through-hole components the backbone of most Maker projects and pre-production prototypes.

There are two types of circuit boards to which through-hole components are commonly mounted mounted: single-sided and multilayer.

Single-sided circuit boards (Figure 4-20) contain copper on only one side of the board and are commonly found in hobbyist projects and low-cost electronics. Having copper on only one side simplifies the design and facilitates easy creation by using the toner-transfer method. Single-sided circuit boards are also significantly easier to rework because the solder only needs to be removed from one surface.

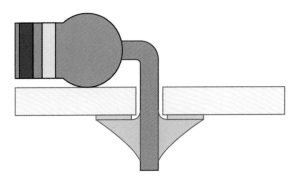

Figure 4-20. *A component soldered to a single-sided circuit board.*

Multilayer boards (Figure 4-21) contain two or more layers of copper attached to the circuit board material. This configuration makes it possible for circuits to take up less surface area because the traces and planes can overlap on the subsequent layers. When soldering through-hole components on multilayer boards, the solder pad typically has two sides: one pad on the top, and one on the bottom of the board. Connecting the layers is a hole plated with conductive material, usually copper or tin. The advantage of using multilayer circuit boards with through-hole components is the added structural integrity provided by the increased surface area.

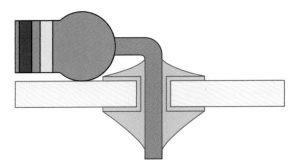

Figure 4-21. *A component soldered to a multilayer circuit board.*

The added thermal mass of pads that connect to each side of the board make components difficult to desolder, especially if they are attached to a ground plane. Often excessive heat applied during this process results in damage to the component or de-lamination of the pad (Figure 4-22).

Figure 4-22. *A scorched circuit board.*

Identifying Poor Through-Hole Solder Joints

Amateurs and skilled solder technicians alike have at some point made improper solder joints or have even damaged their circuit board. This occurs as a result of overheating the solder joint, adding too much or too little solder, or producing a cold solder joint.

Overheated Joint

Overheated joints are the most common mistake made by beginners when learning to solder (Figure 4-23). This occurs when the soldering iron is applied to the joint for too long, resulting in the physical damage of the circuit board or adjoining components. Depending on the severity of the damage, the joint can be corrected by removing any excess solder, letting the circuit board cool, cleaning the area, and fixing the damage with fresh solder. If a pad has been lifted or a trace has been broken, it can be easily fixed with the aid of a jumper wire.

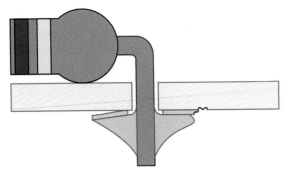

Figure 4-23. *An overheated joint.*

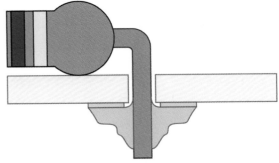

Figure 4-24. *Cold solder joint.*

Use PTFE-coated wire-wrap wire to fix broken traces because the high temperature coating makes the job easy. An alternative is to use the leads of a resistor. Bend the leg at a 90-degree angle in the length that is required and use the resistor as a handle when soldering in place. Trim off the excess lead and the fix is done!

Cold Solder Joint

Cold solder joints occur when the solder is not elevated to a high enough temperature or the joint is disrupted while cooling (Figure 4-24). This results in a drastically weakened joint that will perform unreliably or fail entirely. There are two ways to fix a cold solder joint. First, additional flux can be added to the joint and corrected by simply reapplying the soldering iron again until the joint wets. Second, the joint can be reformed by removing the solder from the joint and applying fresh solder.

Too Little Solder

Applying too little solder to the joint is also a very common mistake (Figure 4-25). This problem isn't as obvious in appearance as it sounds and the resulting joint can actually look correctly formed. The mistake occurs when the pin and pad are unevenly heated. When solder is applied to the joint, it only wets one of the connections rather than both. To fix this problem, reapply heat to the joint so that both the pin and pad are sufficiently heated and then apply a small amount of new solder. The result should be a shiny convex fillet. If the joint looks dull or misshapen, let it cool, apply a small amount of flux, and reapply your soldering iron until the joint has the desired appearance.

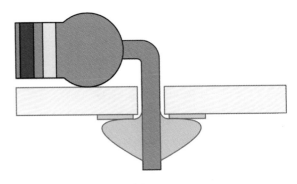

Figure 4-25. *Too little solder.*

Too Much Solder

Applying too much solder is another very common mistake (Figure 4-26). The result is anywhere from a convex blob to a small sphere of solder around the pin and pad. Although this problem does not necessarily affect the performance of the joint, it can lead to electrical shorts with nearby components and doesn't reflect good soldering etiquette. There are two good ways to fix this problem. First, remove the excess solder by using desoldering braid or a vacuum gun, let the joint cool, and then re-apply an appropriate amount of solder. Second, trim the components pin up to the joint, invert the circuit board so that the solder joint is facing the ground, and then carefully apply your soldering iron to the joint. If you are successful, gravity should transfer the excess solder to your iron.

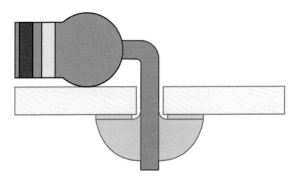

Figure 4-26. *Too much solder.*

Soldering Technique

I started soldering when I was in elementary school, building through-hole projects from schematics found in books and with the help of a cheap, unadjustable soldering iron. This iron provided me with a rather unpleasant soldering experience, leaving my circuit boards riddled with scorch marks and questionable solder joints. It wasn't until I completed four or five circuits that I realized the importance of the relationship between solder, temperature, and time.

Most solder used for soldering electronics contains a small quantity of solder flux. This flux is designed to prepare the contact surfaces by removing impurities and assisting in the proper wetting of the joint. A proper solder joint is

formed by introducing enough heat to the junction between the pin and pad and applying a small amount of solder.

A good rule of thumb when soldering is to complete the joint in under three seconds. This "three-second rule" helps prevent damage to the circuit board and component as well as provides ample time to create a proper joint. If you have not completed soldering the joint in less than three seconds, let the joint cool and then try again. Achieving a properly soldered through-hole joint is not a difficult task. If you understand the variables required to successfully solder a joint and follow the "three-second rule," you should be off and running in no time.

Properly soldered through-hole joints should be concave in shape and should completely wet the pin and pad. These traits illustrate that the joint was made with proper technique and will result in a long-lasting electrical connection.

Materials

Materials List		
Item	**Quantity**	**Source**
Perforated circuit board	1	Electronics supply store
Misc. through-hole components	1+	Electronics supply store

Makerspace Tools and Equipment
Circuit board vise
Safety glasses
Snips
Soldering iron and stand w/ sponge
Solder wire

Proper Through-Hole Soldering

This project requires that safety glasses be worn throughout its entirety.

Step 1

Pass the pins of your through-hole component through their respective holes so that the component is located on the appropriate side of the board (Figure 4-27). Components with long leads, such as resistors, can be held in place by gently bending the leads to the side, or by securing in place with a piece of tape.

Figure 4-27. *Position the soldering iron at the root of the joint, heating both the lead and the pad.*

Step 2

Check whether the soldering iron has reached temperature by applying a small amount of solder to its tip (Figure 4-28). If the solder quickly melts, it has reached temperature, and you can clean the now tinned tip with the sponge. Gently press the iron in the root of the joint so that you evenly heat the pin and pad. Using your free hand, carefully apply enough solder to the joint to sufficiently wet the pin and pad, resulting in a slightly convex fillet. Remember, this entire process should take three seconds or less. If you are soldering a through-hole component to a multilayer circuit board, ensure that the solder completely flows through the plated hole and wets the component side pin and pad with a slightly convex fillet of solder.

Figure 4-28. *Apply enough solder to completely cover the pad and the base of the component's lead. Properly soldered joints have a convex shape and a shiny appearance.*

Project: Simple Constant-Current LED Tester

There are many ways to test whether an LED is functional. You can use a multimeter, a power source, and single voltage-dropping resistor, or with a constant-current power supply. Although each of these accomplish the same general task, it is always nice to have a dedicated piece of hardware. This project (Figure 4-29) illustrates the fundamentals of through-hole soldering through the construction of a constant-current LED tester that can be mounted onto a 9-volt battery with just a couple of components.

Figure 4-29. *The LED tester project.*

This circuit (Figure 4-30) is designed around the popular LM317 adjustable voltage regulator and allows for the quick testing of through-hole LEDs. The LM317 regulator operates by using a voltage feedback loop to regulate output voltage by burning off excess voltage in the form of heat. It can be easily configured as a constant-current power supply that uses its voltage-regulating ability to automatically adjust the output voltage based on current consumption. If desired, the circuit can be modified as an adjustable constant-current supply by adding a potentiometer in series with R1.

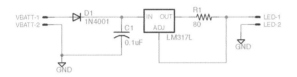

Figure 4-30. *Constant-current LED tester schematic.*

Here's how to calculate the constant-current supply:

(1.25 V / R1) * 1000 = IF

If the LED has an IF (LED Forward Current) of 15 mA:

(1.25 V / R1) * 1000 = 15 mA

or:

1.25V / (15 mA / 1000) = R1

R1 = 83.3 Ohms

For this circuit, the current will be limited to ~15.6mA by using an 80 Ohm resistor.

This project requires that safety glasses be worn throughout its entirety.

Materials

Project Materials List		
Material	**Qty**	**Source**
0.1 uF Capacitor	1	Electronics supply store
1N4001 Rectifier diode	1	Electronics supply store

Project Materials List		
5 mm LED	1	Electronics supply store
~80 Ohm resistor	1	Electronics supply store
9 V Battery	1	Electronics supply store
9 V Battery snap connector	1	Electronics supply store
LM317 (TO92 package)	1	Electronics supply store
Perforated circuit board	1	Electronics supply store
Screw terminal (2 position)	1	Electronics supply store

Project Tools List
Soldering iron
Soldering iron stand
Solder wire
Solder sponge
Wire snips
PCB holder
Double-sided tape
Masking tape

Procedure
Step 1

Start by mounting the perforated circuit board in your circuit board holder, insert the components as shown, and then secure in place with a small piece of masking tape (Figure 4-31). Alternatively, you can carefully bend its legs to the side to prevent the component from falling out. Flip the circuit board over and carefully solder the leads in place. Trim the excess leads when complete.

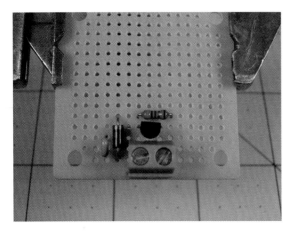

Figure 4-31. *Position the components as shown.*

Step 2

With the solder side up, connect the components together by carefully bridging the circuit board's pads with solder (Figure 4-32). This technique is more reliable than using wires to make the connections, which tend to fatigue and brake with repeated handling. If you accidentally make an unwanted connection, let the board cool for a few seconds and then use the soldering iron to gently remove the unwanted solder.

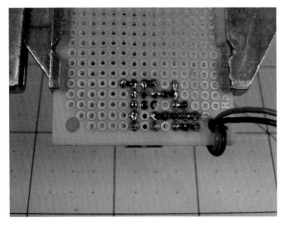

Figure 4-32. *An alternative to making connections with wire is to create pseudotraces by bridging pads with a small amount of solder.*

Step 3

Solder on the battery snap's leads to the illustrated position and mount the battery in place using double-sided tape. Congratulations! You have successfully assembled your LED tester, and now it's time to power it up and test some LEDs!

Troubleshooting

If the power LED doesn't light up when the battery is connected:

- Check that the battery has sufficient voltage.
- Check the orientation of the LED; it might be installed backward.
- Check the orientation of the rectifier diode; it might be installed backward.

If the LM317 gets really hot:

- Check that your input voltage is not over 10 volts.
- Check the orientation of the LM317; it might be installed backward.

The tested LED does not light up:

- Your LED might be bad.
- There might be a short circuit with your soldering. Flip the tester over and inspect the joints and bridges.

Surface-Mount Soldering

As your skills develop and the projects gain complexity, you will find the need for higher levels of soldering skill to properly mount the components. At first, it might seem that surface-mount soldering takes an unobtainable level of discipline and skill. But as with anything, practice makes perfect, and you will soon find that it really isn't any harder than soldering through-hole. The world of surface-mount soldering is exciting and fun and when you get the hang of soldering those tiny components, you will be able to transform your projects into smaller and more robust forms.

Surface-mount, or SMD, components (Figure 4-33) are the primary type of component found in commercial electronic devices. As electronics become more efficient, produce less heat, require less voltage and as manufacturing capability improves, the size of components decrease. This saves both materials and cost while simultaneously affording higher-density layouts. A good example of this forward progression is illustrated when comparing the early room-sized computers of the 1970s to the multicore-powered phones of today.

Figure 4-33. *A properly soldered SMD component.*

Each SMD component contains a series of small leads or contacts that are the electrical connection to the circuit board. This design differs from through-hole components because they tend to take up more space and require their leads to pass through the circuit board. Surface mount components are purchased on reels or as cut-tape (a cut section of a reel). This format makes it possible to feed surface-mount components into machines that pick and place the components onto prepared circuit boards using sophisticated imaging and control technology.

> *Make sure to thoroughly read the specification sheet for your components because their power rating often decreases with size. You don't want to use a component that isn't properly rated; it is likely to fail or might even catch fire.*

As projects are refined, it often makes sense to convert the design to use surface-mount components. The sizes of these components vary greatly, so when you are just starting out, go with the "big" ones. Table 4-6 presents some common components that are good for beginners and their package sizes.

Table 4-6. Recommended beginner SMD component sizes

Package	Component Types	Lead Spacing (in)
0805	Resistors, ceramic capacitors, LEDs	0.080
1206	Ceramic capacitors, diodes, shunts	0.120
SM-A	Tantalum capacitors	0.126
DPAK	Voltage regulators, power diodes	0.180
SOIC	Microcontrollers, EEPROMs and other general integrated circuits	0.050

As your skills increase, you will find that you can handle even smaller components. This will lay the groundwork toward making smaller and more dense layouts in addition to reducing component and material cost. Table 4-7 lists some common components and their package sizes that are good for more advanced technicians.

Table 4-7. Recommended advanced SMD component sizes

Package	Component Types	Lead Spacing (in)
0603	Resistors, ceramic capacitors, LEDs	0.060
SOT23	Transistors, diodes	0.075
SOT23-5/6	Voltage regulators, microcontrollers	0.037
SSOP	Microcontrollers, EEPROMs and other general integrated circuits	0.025
QFP	Microcontrollers	0.026

Each of the aforementioned package types require a slightly different method for proper soldering. You can solder some components by hand with the assistance of one or two soldering irons, whereas others, like the BGA package, require more complex methods.

Identifying Poor SMD Solder Joints

The occurrence of improper solder joints is even more prevalent with SMD components. These problems often go unnoticed due to the joint's small size and obscurity, but they are identical to those found when soldering through-hole. Unfortunately, they are also more complicated to resolve. With your magnifying glass in one hand

and iron in the other, you can easily identify and eradicate them.

Cold Solder Joint

The occurrence of cold solder joints on SMDs is relatively low due to the heavy use of flux (Figure 4-34). Even still, there are times when a joint doesn't properly form and rework is necessary. To fix the problem simply apply a small amount of flux to the joint and resolder.

Figure 4-34. *SMD with a cold solder joint.*

Overheated Joint

Many surface mount components feature pins and pads that are a fraction of a millimeter wide (Figure 4-35). This small amount of material is easily overheated and damaged if the soldering iron drags across the surface. These joints can be fixed by removing the excess solder with desoldering braid and clearing away any circuit material by using alcohol. Continue to solder the rest of the component's pins, and when secure, repair the broken pin/pad using a small piece of wire-wrap wire.

Figure 4-35. *Overheated SMD.*

Too Little Solder

Solder joints with too little solder are hard to detect on through-hole components and are virtually impossible to detect on SMDs without magnification (Figure 4-36). These joints occur when the solder does not properly wet the junction between the pin and the pad, causing a small gap to form. To rem-

edy the problem simply apply a small amount of flux to the joint and solder it again. Alternatively, you can fix the problem by gently pressing down on the pin with the soldering iron until the pin and pad properly join.

Figure 4-36. *SMD with too little solder.*

Too Much Solder

It is very easy to create solder bridges when soldering SMDs. Happily, there are two good ways to solve this problem (Figure 4-37). First, continue to solder the remaining pins to secure the part and, when secure, apply a small amount of flux to the solder bridge and wick it away with the tip of a clean soldering iron. Because the soldering iron likes to bond to solder, it will suck up the tiny amount of solder left on the pins. The second method utilizes conventional soldering braid. This method is recommended as a last resort only because the mechanical contact of the braid and soldering iron can damage the pins. To remove excess solder with the braid, start by tinning it with a small amount of solder and lay it on top of the pins. Using the side of the soldering iron tip, gently press down on the braid until it wicks away the unwanted solder.

Figure 4-37. *SMD with too much solder.*

Hand-Soldering SMDs

I learned my technique for soldering SMD components from a technician who makes her living by hand-soldering $75,000 computer chips for the space industry. This is quite a feat to witness, and her ability to do so with gracefully steady hands is awe inspiring. Even though she has multiple thousand-dollar soldering irons and a fleet of tips at her disposal, she is able to accomplish most of her tasks with just two irons outfitted with her favorite tip. The reality is, you do not need multiple rooms full of thousand-dollar irons to begin surface-mount soldering. A pair of adjustable irons costing less than $100 can provide plenty of capability and doesn't strain the budget.

This section focuses on two soldering iron techniques: single-iron and dual-iron. Single-iron technique is good for attaching new components to untinned pads because you rely on the absence of solder on one side to keep the component flat while soldering. Dual-iron technique works well for all components with two or more pins and aids in reworking components.

Set the iron's temperature to 371°C/ 700°F and adjust slightly, depending on the type of part you are soldering. If I am soldering a joint that is coupled to the ground plane, I bump the temperature up a bit to accommodate for the added thermal mass, and vice versa. Also, don't forget the "three-second rule" ("Soldering Technique" on page 105)!

Materials

Materials List		
Item	**Quantity**	**Source**
Test SMD circuit board	1	Electronics supply store
0805, SOIC, and DFN SMD Components	1+	Electronics supply store

Makerspace Tools and Equipment
0.015 in diameter solder wire
Adjustable soldering irons w/ chisel tips
Circuit board vise

Makerspace Tools and Equipment
Safety glasses
Snips
Soldering iron stands w/ sponges
Tweezers

Single-Iron Technique: 0805 Package

This project requires that safety glasses be worn throughout its entirety.

Step 1

Secure your board to the work surface. A soldering vise comes in handy here, but work with what you have. The first component will be a 0805 resistor. These resistors are nice to work with due to their relatively large size and lack of polarity.

Step 2

Apply a small amount of solder to one of the pads (Figure 4-38). If you have the choice between a ground or nongrounded pad, choose the nongrounded because they tend to sink a lot of heat. Carefully pick up the resistor with your tweezers and position it in line with the solder pads. With you free hand, heat up the solder and the resistor's contact until they are wet.

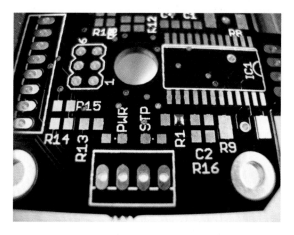

Figure 4-38. *Apply a small bead of solder to one of the 0805 pads.*

Step 3

Using a small amount of fresh solder, solder the second pad (Figure 4-39). Apply a small amount of solder to the first joint and sweep away the excess. This will help to produce a nice, shiny joint (Figure 4-40). Optionally, you can add flux to the joint and touch up with the soldering iron.

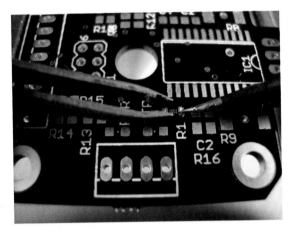

Figure 4-39. *Hold the component in place with tweezers while soldering the pad. If the component shifts during the process, let it cool before reworking.*

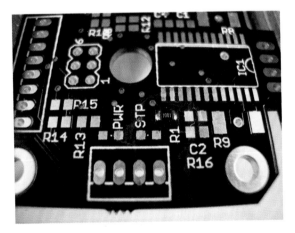

Figure 4-40. *Visually inspect the joints to ensure that each is properly connected.*

Single-Iron Technique: SOIC Package

This project requires that safety glasses be worn throughout its entirety.

Step 1

The next component will be an SOIC package EEPROM (Figure 4-41). This package type is nice because the leads have relatively large spacing. Apply a small amount of solder to one of the pads. Use pads on the corners and make sure that it is nongrounded. Upon completion, apply a small amount of flux to the remaining pads. This will facilitate quicker wetting of the remaining joints.

Figure 4-41. *Apply a small bead of solder to one of the SOIC pads.*

Step 2

Using the tweezers, position the component above the respective pads, observing the pin location indicator. Heat up the lead and pad with the iron until they bond, ensuring that the component is properly aligned. If it is not, reheat the joint and align accordingly.

Step 3

Solder a lead on the opposite side of the component by placing the soldering iron to the side of the lead so that it touches both the lead and the pad (Figure 4-42). Then, add a small amount of solder to the opposite side of the lead until the solder sufficiently wets the joint. Be careful not to heat the adjoining lead so that you avoid the risk of bridging them with solder. Soldering this opposing lead helps to hold the component in place while you solder the rest of the leads. Don't worry if the first two joints are not perfect; you will be cleaning them up at the end. Carefully solder the remaining leads and clean up any poorly made joints after applying a small amount of flux.

Figure 4-42. *Visually inspect the joints as you go to ensure they are properly formed and connected.*

Single-Iron Technique: DFN Package

This project requires that safety glasses be worn throughout its entirety.

Step 1

The final component you'll solder will be a DFN (Figure 4-43). This package is designed to be soldered via the reflow process and often has an exposed pad on the bottom of the component that requires some fancy footwork to solder. These pads are usually connected to ground and can sometimes be excluded. But check your datasheet just in case! Start by adding a small amount of solder to one of the nongrounded corner pads.

Figure 4-43. *Apply a small bead of solder to one of the DFN pads.*

Step 2

Upon completion, apply a small amount of flux to the remaining pads (Figure 4-44). Using tweezers, hold the component in place above the respective pads, observing the pin location indicator. Carefully heat up the pin and pad until the solder fully wets the joint. Continue to solder the remaining joints, beginning on the opposite side of the component to prevent overheating.

Figure 4-44. *Visually inspect the joints to ensure that each is properly connected.*

If you are having trouble getting the contacts on a QFN or DFN to sufficiently wet, try dragging a ball of solder across the contacts after you have applied a new layer of flux (Figure 4-45). The surface tension of the solder ball should prevent bridges and encourages an even distribution of solder. Just be careful you don't overheat the part!

Figure 4-45. *Dragging a small ball of solder across the pins and pads can make the soldering job easier. Just make sure you use a good amount of flux.*

Dual-Iron Technique

This project requires that safety glasses be worn throughout its entirety.

Step 1

The first component will be the same type of 0805 resistor used with the single-iron method (Figure 4-46). Begin by applying a small amount of solder to both of the pads. Ensure that there is an even amount of solder deposited. If it is uneven, quickly swipe the tip of the soldering iron across the pads to wick away any excess solder. Place the resistor close to the prepared pads for easy access.

Figure 4-46. *Add a small amount of solder to both pads prior to soldering the component.*

Step 2

Using the tips of the soldering iron, quickly lift the resistor and place on top of the prepared pads (Figure 4-47). During this action, the solder on the pads should reheat and wet the resistor's contacts. Align the component and remove the irons simultaneously. If you find that the part shifts when you remove the irons, you can try removing one at a time. Often times one side will solidify before the other as a result of inconsistent temperature. Apply a small amount of flux to the component and clean up the joints with a quick swipe of the soldering iron.

Figure 4-47. *Place the component using both irons.*

Reflow Soldering SMD Components

Hand-soldering surface-mount components is a great way to make small projects with relatively high complexity. But, there might come a time when you will need to make higher quantities of boards or have the need to use component packages with contacts that are physically impossible to hand solder. In this case, the reflow soldering method is the way to go. This method utilizes sophisticated equipment and processes to deposit a conformal layer of solder onto one or both sides of a circuit board. It then systematically places and aligns all of the desired components. The freshly populated board is then placed in a temperature-controlled oven in which the board is taken through a series of thermal cycles that are designed to properly wet the solder joints and burn off any impurities. The result is a populated circuit board with highly consistent solder joints.

Reflow Soldering Equipment

The Solder Stencil

The solder stencil is responsible for masking the areas around each solder pad and making it possible for a small layer of solder paste to be deposited onto the pad (Figure 4-48). The quantity of solder is dependent on the size of the cutout and the thickness of the stencil. Typically, hobbyist stencils are made out of mylar because it can be cut on a laser cutter. This material accommodates the proccessing of multiple panels of boards before any noticeable damage to the stencil occurs. In industry, stencils are made out of thin stainless steel. This sheet is etched in a process similar to the photo-lithographic method used to manufacture circuit boards or can even be cut with laser. The advantage to metal stencils over mylar is the precision of the cutouts, with metal stencils providing a much more accurate deposition of solder paste in addition to the added durability.

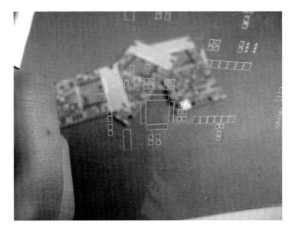

Figure 4-48. *A solder stencil.*

You can purchase inexpensive mylar solder stencils online at retailers like Pololu (http://www.pololu.com/) and OHARARP LLC (http://ohararp.com/).

Solder Paste

Solder paste is essential to the reflow process (Figure 4-49). Depending on the type of components and environmental conditions, there are a number of types of solder paste from which you can choose. Choose a paste with the alloy and flux type required for your project.

Figure 4-49. *Solder paste.*

Reflow Oven/Plate

The reflow oven is the most critical component to the reflow soldering process (Figure 4-50). This oven acts to evenly heat the circuit board, solder, and components through a highly controlled thermal cycle following what is known as a *reflow profile*. This reflow profile has a series of thermal plateaus that are designed to properly burn off any unnecessary additives in the solder prior to wetting and evenly heat the joints. After a short period, the oven elevates the temperature to the point that the solder sufficiently wets, and then gradually lowers the temperature until the joint has properly solidified. Unlike the hand-soldering process, reflow is designed to solder multiple boards at once as well as boards that are designed with components on both sides.

The following procedure for the reflow soldering method assumes that you have all of the required equipment. If you do not, this chapter also provides direction for constructing a reflow oven based on a household electric skillet.

Figure 4-50. *Skillet reflow.*

Each solder paste manufacturer has a profile optimized for their solder. A components may also have profiles that are designed around that component's temperature range.

Materials

Materials List		
Item	**Quantity**	**Source**
Test SMD circuit board w/ components	1	Electronics supply store

Makerspace Tools and Equipment
Drafting triangle or equivalent 90-degree reference
Glass sheet
Isopropyl alcohol (>90%)
Masking tape
Paper towels
Reflow oven or skillet
Rulers
Safety glasses
Solder paste
Squeegee
Tweezers

Reflow Soldering Process

This project requires that safety glasses be worn throughout its entirety.

Step 1

Start by compiling all of the tools and materials necessary for this project on the same table and within close proximity to one another (Figure 4-51). The solder paste has a relatively short time frame in which components can be held in place via the solder's tack. As soon as the the volatiles evaporate, it is much harder to prevent the components from being bumped off of their respective pads. It is OK if you do not perfectly align the components prior to putting them in the oven because the surface tension of the molten solder will help to realign any mistakes.

Figure 4-51. *Plastic putty knives used for spackling work well for spreading solder paste. Try a few different types (plastic, rubber, and metal) and decide which works best for you.*

Step 2

Secure the two rulers to the glass sheet perpendicular to each other by using a small amount of masking tape. Ensure that they meet at a 90-degree angle. Accuracy in this step is critical to guaranteeing the even distribution of solder paste over multiple boards. Evenly align a series of bare circuit boards on the glass sheet so that they are oriented in the same direction as your solder stencil. Gently place your solder stencil on top of the circuit boards so that the cutouts align with their respective pads. Once the stencil is aligned, tape only the top of the stencil to the glass sheet. This will allow you to apply paste to multiple panels of boards without having to realign the stencil. After everything is secure and the stencil is positioned, deposit a small bead of solder paste along the top of the stencil.

Step 3

Holding the bottom of the stencil with one hand and your squeegee in the other, wipe the solder down the stencil, applying even pressure and speed (Figure 4-52). The objective on the first wipe is to ensure that all of the cutouts have been filled with solder paste. Be careful not to shift the stencil during this process because it will result in im-

proper deposition of solder onto the circuit board. If this occurs, scoop up the solder paste with your squeegee, wipe the boards and stencil clean with alcohol, and try again. Continue to hold the stencil in place and again pass the squeegee over the stencil. This removes any excess solder paste and ensures that every pad is evenly coated. If you shift the stencil during this process, you must wipe the boards clean and begin again.

Figure 4-52. *Apply constant and even pressure while you drag the solder paste across the surface of the stencil.*

Step 4

Carefully peel up the stencil in one smooth motion so as to expose the freshly prepped circuit boards. Remove the circuit boards from the fixture and place them on your work surface, being careful not to disturb the solder paste. Visually inspect each board for unwanted solder deposits or bridges. If you find one, use a tooth pick to correct the problem. The final product can have a little bleeding or overlap because it will be corrected during the heating process.

Step 5

Preheat your oven using the recommended profile and align each board so that they can be easily accessed (Figure 4-53). Populate the boards with the appropriate components, using tweezers or your fingers; just be careful not to smear the solder. Verify that components with position indicators such as diodes, polar capacitors, and integrated circuits are positioned properly. When the oven has reached the designated start temperature, carefully pick up and place your circuit boards onto the heated surface. Activate the solder profile and watch in amazement as the solder joints form before your eyes. If you are manually controlling your oven, use a timer for each stage of the profile. There is a short period of time during the reflow stage during which you can make minor adjustments to the components. Do this using the utmost care because the neighboring boards are very easily disturbed.

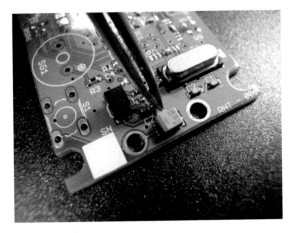

Figure 4-53. *Depending on your setup, you can fine tune the position of the components during the "soak" portion of the reflow profile.*

Step 6

Wait until the oven has cooled before removing your boards. The circuit boards should now be completely soldered! Inspect each board for misaligned parts, bridges, and unused solder. Wasn't that easy?

Project: Skillet Reflow

There are times when hand-soldering SMD components becomes too laborious or even impossible. It just so happens that a quick trip to the local home goods store can provide reflow equipment for under $100. The reflow process makes possible the simultaneous soldering of all of the components on a circuit board, and can even be done on both sides of a circuit board.

This project requires that safety glasses be worn throughout its entirety.

This project involves components that operate on household AC power. Never attempt to assemble or disassemble this project while they are plugged in.

Materials

Materials List		
Item	**Quantity**	**Source**
1/4-20 × 1 in bolt	1	Home improvement store
1/4 in flat washer	2	Home improvement store
1/4 in Nut	1	Home improvement store
1200W Electric skillet	1	Home goods store
22 gauge stranded sire	1	Electronics supply store
6-32 × 1 in bolt	2	Home improvement store
#6 washer	2	home improvement store

Materials List		
6-32 nut	2	Home improvement store
Adhesive cable tie bases	6	Electronics supply store
Electronics enclosure	1	Electronics supply store
Computer power cable	1	Computer supply store
PID temperature controller	1	Electronics supply store
PowerSwitch Tail II	1	Electronics supply store
Thermocouple	1	Electronics supply store

Makerspace Tools and Equipment
Wire stripper
Screwdriver
Drill with 1/4 in bit
Safety glasses
Rotary tool with cut-off wheel
Cable ties

Procedure

Step 1

Begin by determining the location of the components inside of the electronics enclosure. Mark the location of the PowerSwitch Tail's mounting holes and the rectangular cutout for the PID as directed by your PID's manual. An old rack-mount computer case is used in this example. Cut out the rectangle for the PID by using a rotary tool with a cut-off wheel.

Step 2

Unscrew the base of the electric skillet and mark a location for the temperature sensor 3/8 in away from one side of the heating element. Drill a 1/4 in hole at the location, taking care not to damage the heating element. Secure the temperature sensor in place with the bolt by sliding one washer over the threads, pass it through the hole, and then secure in place with another washer and the nut. Position the skillet onto the side of the enclosure and mark the mounting locations. Drill holes for each foot and attach the skillet to the side of the enclosure.

Step 3

Drill the PowerSwitch Tail's mounting holes and secure in place with the 6-32 bolts followed by washers and nuts (Figure 4-54). Mount the PID in the cutout. Cut and strip two lengths of the stranded wire and connect them to the PID's control output. Strip and connect the other ends of the wires to their respective locations on the Power-Switch Tail. Verify the polarity. Connect the temperature sensor to the PID, also accounting for polarity.

Step 4

Ensure that the computer power cable is **not plugged in** and cut off the C13 connector. Strip off 2 in of the outer insulation and the ends of the three wires. Connect the wires to their designated locations noted in the PID manual. Pass the power cable out of the enclosure but **do not plug it in**!

Step 5

Plug the electric skillet into the PowerSwitch Tail and set the temperature to MAX (Figure 4-55). Coil the remaining cable and secure with a tie. Secure any loose wires by using the cable tie mounts and ties and check for any loose wires. Close up the enclosure and plug in the PID. If everything is hooked up correctly, your PID should illuminate showing you about 20° C (or whatever your room temperature is). Configure your PID according to the manual and, when ready, plug in the PowerSwitch Tail. The controller should now power on and off the skillet as it reaches the desired temperature. Remember, only use this skillet in a well-ventilated area. The fumes released during reflow can be harmful, and every precaution should be taken to ensure the safety of you and those around you. Congratulations! You have just assembled a reflow skillet that will serve your Makerspace for years to come!

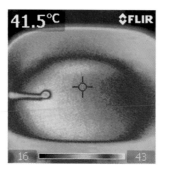

Figure 4-55. *Some electric skillets have "hot spots" which are located through trial and error. Typically the best location is directly above the element and with the lid on.*

Troubleshooting

If the PID does not turn on:

- Disconnect the power from the source and ensure that the leads have been appropriately connected to the PID.

If the PID turns on, but the skillet does not heat up:

- Verify that the temperature knob on the skillet is set to its highest point.
- Ensure that the PID is properly connected to the PowerSwitch Tail.

If the skillet heats up, but the solder never melts:

- Some skillets are designed not to reach 500° F due to thermal degradation of the skillet's coating. Purchase a different skillet and **do not** exceed the skillet's maximum temperature, because you run the risk of damaging your components as well as releasing toxins from the skillet's non-stick coating.

Desoldering and Rework

It's going to happen: you just discovered that you reversed the polarity on a capacitor or your surface-mount integrated circuit was installed backward; bummer man! Rework or desolder

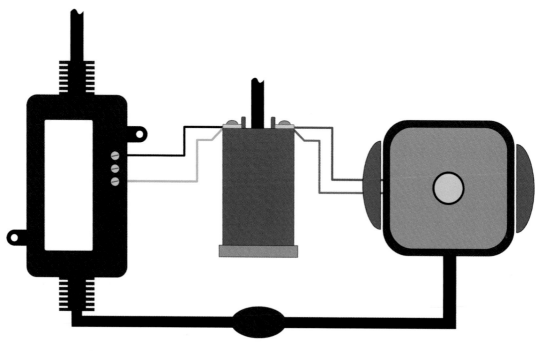

Figure 4-54. *Connect the components according to the diagram.*

that sucker and get your project up and running. Rework is a great skill to learn and can save you save a lot of money in components by scavenging them from unused circuitry.

Reworking and desoldering are the skills that, when mastered, will allow you to better troubleshoot and repair errors on current projects as well as salvage components from existing boards, ultimately saving you time and money. The following section provides descriptions and methods for using some of the common tools employed when reworking and desoldering.

Desoldering Tools

Desoldering Braid

Desoldering braid comes as a spool of tightly woven copper strands impregnated with a small amount of flux (Figure 4-56). This braid is used in conjunction with a soldering iron to remove solder through capillary action and the copper's natural wettability. You use desoldering braid by first placing the braid on top of a solder joint and gently applying the soldering iron. The iron then heats both the braid and the solder until the solder is wicked away from the circuit board. This tool is useful for cleaning pads after components have been removed, and although you can use it to remove through-hole components, it is often difficult and can result in damage to the circuit board due to overheating.

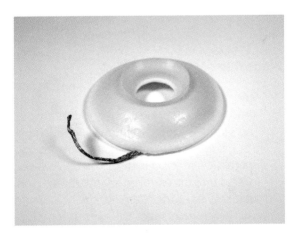

Figure 4-56. *Desoldering braid.*

Apply a small amount of solder to the braid prior to desoldering. This helps accelerate the heating process and therefore reduces the amount of time in contact with the joint.

Solder Sucker

A solder sucker or vacuum is a syringe-like device that uses a button activated piston to suck away molten solder (Figure 4-57). You use this tool by first heating the solder joint with a soldering iron until the solder is completely molten, and then quickly covering the joint with the tip of the solder sucker and pressing the button. Most of the solder will be sucked into the body of the tool and can be removed when necessary. The solder sucker is primarily used for removing solder from through-hole solder joints because it is capable of removing a large quantity of solder quickly.

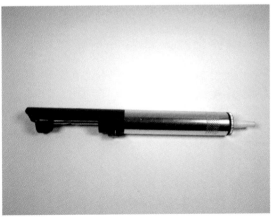

Figure 4-57. *Solder sucker.*

Specialty Solder

There are solder alloys specially designed for the desoldering and rework process (Figure 4-58). These alloys have a much lower melting temperature than standard solder and you mainly use them for the safe removal of SMD components. Working in conjunction with a flux formulated for this process, solder is applied with a soldering iron to all of the pins that secure the component in place. When every pin is covered, you can lift the component with tweezers or a suction cup. If you do this quickly enough, all of the solder can be removed by giving the component one gentle tap on your work surface. Alternatively, you can remove the solder by laying one row of pins along a piece of desoldering braid, which wicks it away with a quick pass of the soldering iron.

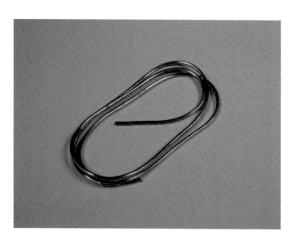

Figure 4-58. *Specialty desoldering solder.*

Project: Inexpensive Hot-Air Desoldering/Rework Tool

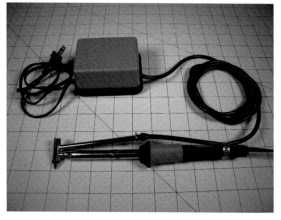

There's not much that's more frustrating than installing a component backward. Especially if it is a fine-pitch device with a lot of pins. The easiest way to remove or replace a damaged or incorrectly installed component is with hot air. Although this method is a lot less invasive than using a soldering iron, it requires expensive equipment to complete. This project is designed to show you how to convert an inexpensive bulb-type desoldering iron into a hot-air gun for desoldering and rework.

This project requires that safety glasses be worn throughout its entirety.

---

*This project involves tools that operate at exceedingly high temperatures. Use the utmost care when working with these tools and **never** rest the iron on a flammable surface.*

Materials

Project Materials List		
Material	**Qty**	**Source**
1 in hose clamp	1	Home improvement store
1/8 in × 36 in silicon tubing	1	Hobby store
45 W desoldering iron with bulb	1	Electronics supply store
Diaphragm-type aquarium air pump	1	Fish supply store
Small piece of brass wool	1	Electronics supply store

Project Tools List
9 mm socket or open-ended wrench
Scissors

Procedure
Step 1

Start by using the 9 mm wrench and carefully remove the soldering iron's tip (Figure 4-59). This exposes a small section of inner chamber. Using the pair of scissors, cut off approximately 1/4 in × 1/4 in of the brass wool and insert it into the exposed chamber. This wool increases the surface area of the heater and will therefore increase the output temperature. Carefully re-attach the tip with the wrench and remove the rubber bulb from the back of the air tube.

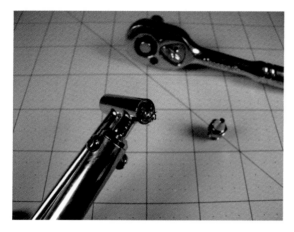

Figure 4-59. *Ensure that the brass wool does not clog the air inlet.*

Step 2

Slide approximately 1/2 in of the tubing over the end of the air tube. Carefully open the 1 in hose clamp and slide over the iron's handle and air tube. Rewind the hose clamp so that the screw mechanism is positioned on top of the tube. You can use this clamp to adjust the airflow by gently loosening and tightening the screw.

Step 3

Finally, attach the free end of the aquarium tubing to the air pump and the project is complete! Congratulations! You have successfully constructed a hot-air desoldering and rework tool for under $20!

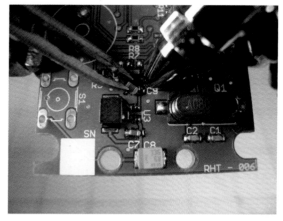

Figure 4-60. *Position the tip of the iron about 0.1 in above the part and move in a circular pattern until the solder melts.*

Troubleshooting

If the iron does not heat up:

- Make sure it is plugged in.

If the iron does not blow hot air:

- Check the air hose connection from the pump to the air tube for kinks, or that your hose clamp is tightened too tightly.
- Check that the brass wool is not constricting the airflow.

If the hot air is not hot enough to melt the solder:

- Check for airflow problems. Sometimes, if the airflow is too high, it doesn't have time to heat up. Alternatively, if the airflow is too low, not enough hot air is produced to heat the solder joint.
- Move the tip closer to your project.
- On stubborn parts, you can preheat the board with a hair dryer or heat gun.

Playing with Electronics | 5

IN THIS CHAPTER

- Facilitating Design
- Soft Circuits
- Making Circuit Boards
- Working with Microcontrollers

Most Makerspaces can accommodate electronic projects that range from fuzzy soft-circuits to complex surface-mount monstrosities. Regardless of the complexity, the starting point of any electronics project (Figure 5-1) should be to visualize how the project works. I prefer to use a computer to create a virtual circuit prior to prototyping because it helps me iron out potential problems. How you approach the task is up to you, but remember, "prior planning prevents poor performance!"

Figure 5-1. *A microcontroller-powered project.*

A Makerspace should contain all of the equipment necessary to facilitate the reference, testing, and modification of new and existing electronics projects. This includes tools like the multimeter, oscilloscope, solder and a soldering iron, test leads, wire, a computer, and most important, available power outlets. The most common of these tools are illustrated in Table 5-1:

Table 5-1. Electronics design and testing equipment

Tool	Function
Adjustable-temperature soldering iron	Simplifies the soldering process
Adjustable power supply	For powering projects at their required voltage and current
Computer	For reference resources and design
ESD safe mat	Protects electronic projects from electrostatic damage
Isopropyl alcohol	For cleaning up after soldering
Lamp w/ magnifier	For soldering and fine adjustments
Oscilloscope	Electronics testing and evaluation
Parts storage	Outfitted with various common useful components
PCB vise	Isolates PCB while soldering and tuning
Power strip	For easy access to outlet power
Solder spools	Different diameters for through-hole and surface-mount device (SMD) soldering
Trash can	Helps keeps the bench clean and tidy
Wire	22-gauge solid core wire for breadboarding and stranded for prototyping

Facilitating Design

Start small. Regardless of the size of your project, the key to creating a successful circuit is to break it down into smaller chunks you can test and verify more easily. Then, gradually piece the chunks together until the circuit is complete. Some like to start this process with software design and simulation; others go straight to the breadboard. Ultimately, the goal is the same: produce a functional electronic project. The resources in Table 5-2 provide an excellent source for in-depth information ranging from basic electronics to complex electrical engineering.

Table 5-2. Resources of basic electronics and design

Title	Author	Overview
Getting Started with Arduino (2008 O'Reilly)	Massimo Banzi	An introductory overview of the popular Arduino development board.
Make: Electronics (2009 O'Reilly)	Charles Platt	A hands-on approach to learning basic electronics.
Make: Wearable and Flexible Electronics (2013 O'Reilly)	Kate Hartman	Materials, tools, and techniques for soft-circuit electronics.
The Art of Electronics (1989 Cambridge University Press)	Paul Horowitz/ Winfield Hill	Comprehensive "go-to" resource for electronic concepts and design.

Finding Parts and Understanding Datasheets

As you spend more and more time working with Maker technology, you will begin to see recurring components and designs. Parts like FTDI's FT232RL has emerged as the staple USB-to-Serial converter integrated circuit (IC) and you can find it in numerous Arduino and adapter designs. Making these common components available for use in the Makerspace helps to spur innovation and acts as an invaluable resource when components fail.

Keep components organized in bins or drawers rather than bags, as shown in Figure 5-2. This not only helps keep clutter to a minimum, but helps keep track of inventory. Small components like

SMDs can be stored in fishing tackle boxes, or fit nicely into trading-card binder sheets.

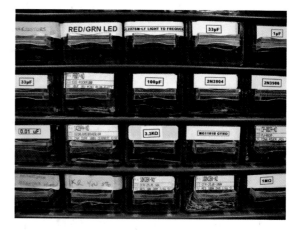

Figure 5-2. *Organization matters.*

Some components can even be obtained as a sample from the manufacturer. Some manufacturers of these components are interested in being integrated into your design and help you with your prototyping by offering small quantities of components for free. If you choose to request samples on behalf of your Makerspace, make sure you do so with moderation and without falsifying information. This service is complimentary and can therefore be easily taken away.

Ladyada.net maintains a useful database of IC manufacturers that support samples (http://www.ladyada.net/library/procure/samples.html).

Component datasheets are provided by manufacturers to offer a detailed overview of a component's feature set and functional requirements. These datasheets are often difficult to follow and can contain far more information than you might require to make an informed decision. Each electronic component datasheet contains a series of common sections; when you become familiar with them, you will be able to quickly pinpoint the desired specification.

Keep a binder of all the Makerspace's common component datasheets as a quick "go-to" reference for designing and hacking.

Datasheet Sections

Features

This section is most commonly located at the beginning of the datasheet. Manufacturers use this to illustrate the component's primary functions. These might include operating voltages, bit rates, input/output types, interfaces, and temperature ranges.

Component Description

This section contains mostly information related to the benefits of the product and helps to explain the listed functions in the features section. Although this section can be useful for answering initial component usage questions, most of this information doesn't help with the decision of whether this component will be useful.

Pin Definition

The pin definition contains diagrams of all of the package types in which the component is available, and you can use it to determine the component name's sub-type, which is used for purchasing. Each package diagram contains a pin number and a label that illustrates that pin's function. These functions are described in greater detail in a "pin function table" that details each pin number, its label, and a brief description of its function.

Electrical Characteristics

This section is broken into two parts: Absolute Maximum Ratings, and Electrical Specifications. The maximum ratings table details the environmental limitations of the component. This table is quite useful, especially when working with components operating at voltages other than that of your microcontroller. It contains information including maximum voltage, current limits, and temperature ranges. The second section details all of the electrical characteristics the component exhibits in all of its modes of operation. It includes voltage levels, current draws, frequencies, sample rates, as well any other pertinent information.

Component Operation

The component operation section includes detailed text describing the component's use and operation. Embedded in this text will often be example code and circuits that can be exceedingly useful, especially when prototyping a circuit.

Packaging Information

The final section details the dimensions of the available package types. This section comes in handy for developing custom component libraries in your circuit-design software.

Computer-Based Circuit Design

CadSoft's EAGLE PCB design software has quickly become the application of choice for circuit and PCB design for the Maker community. With its "Lite" and "Freemium" versions, anyone can begin turning their designs into high-quality circuit boards. After a little practice and , you should be able to quickly turn out designs that can both be produced by hand or with a professional manufacturer. You don't have EAGLE? Not a problem; Table 5-3 presents three other software suites that have similar functionity and will get your project up and running in no time.

Table 5-3. Freely available EAGLE alternatives

Software Title	Developer	Platform	Link
KiCAD	KiCAD developers team	Windows/ Linux	http://www.kicad-pcb.org/display/KICAD/KiCad+EDA+Software+Suite
Fritzing	UASP Interaction Design Lab and community contributors	Windows/ OS X/Linux	http://fritzing.org/
gEDA	OS X/Linux	Ales Hvezda and gEDA Developer Team	http://www.gpleda.org/

Offering the capability to design and manufacture circuit boards greatly increases your Makerspace's electronics prowess and ability to focus more on in-house design. Armed with a base understanding of circuit design and board layout, your Makerspace will be churning out boards in no time.

Working with EAGLE Libraries

Before a schematic takes shape, verify that you have parts libraries for every component in the design. Although you can create parts as you go, having a known list of available parts in your library makes the design process go much more smoothly. Creating parts for a library is also a great project for a Makerspace. As the Makers become more and more familiar with the parts they regularly use, each of these components can be added to your Makerspace's own parts library!

EAGLE comes preinstalled with a vast database of parts that you can access from the Control Panel as well as from within the Schematic and Board layout tools. Each device in the library contains three parts, the symbol, the package, and the device itself. The Symbol dictates how the device looks when it is inserted into the schematic and contains pins with all of the available connections. The Package illustrates how the device looks when it is added to the board layout and transfers the connections made on the sche-

matic to the pads on the device. The Device is the merger between the schematic and the board. It illustrates which pin on the symbol connects to which pin on the package.

The list of available libraries can be viewed by going to the Control Panel and clicking the "+" character next to Library. Each library file, indicated by the *.lbr* file extension, is filled with all of the available devices and packages. All of these files can be individually edited, which comes in handy when repurposing packages for new parts.

Many Makerspace-friendly web-shops like Adafruit and SparkFun have created EAGLE libraries to support the products they sell. You can find more information and files at Adafruit (http://bit.ly/16dO793) SparkFun (http://bit.ly/1bU5FN1).

To open a library for editing, right-click the filename and select Open. Alternatively, you can open the library directly from the toolbar by clicking File→Open→Library. The library editor contains all of the tools necessary to edit the packages and devices, or add new content. Each library consists of three components: the package, the symbol, and the device.

Eagle Library Components
Package

Packages (Figure 5-3) are drawings that illustrate the circuit board mounting configuration for each part type. Parts may be available in multiple package styles, which can be reflected in the part editor. Packages should reflect all of the electrical connections supported by the part and should be named using a representative convention (P1, P2, GND, etc.)

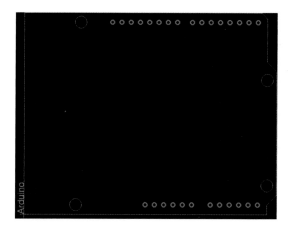

Figure 5-3. *Eagle package.*

Symbol

Symbols (Figure 5-4) are used in the schematic design program and illustrate all of the part's connections. The names and placement of the pins should reflect those found on the actual package and should be named using a convention that illustrates the pin's function (VCC, GND, PB1, PB2, etc.).

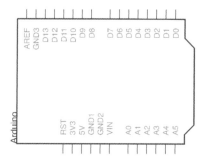

Figure 5-4. *Eagle symbol.*

Device

Devices consist of a series of packages and symbols that contains all of the electrical connection information to connect each pin on the symbol to the pad on the package. Devices can also be configured as different variants that accommodate for that device's available package types.

Project: Creating a Custom Library in EAGLE

There are times when the base EAGLE library does not contain the part you need. To use your desired part within the schematic and board layout tools, you need to add it to the library. This is relatively simple to do, and when you master it you will greatly expedite the design-to-reality process.

Required Software	
Software	**Source**
EAGLE PCB v6 or newer	*http://www.cadsoftusa.com*

Create the Device
Step 1

To begin creating a new library file, in the Control Panel, click File→New→Library. EAGLE creates a new blank library named *untitled.lbr* and places it in the software's default location, *~\EAGLE-6.2.0\lbr*. Before you proceed, click File→Save and save your library as *Arduino.lbr*. This library will contain a through-hole Arduino designed around the ATmega8/168/328 processor and will involve the creation of the package as well as the symbol and the part. Alternatively, you could use the standard ATmega8 device from the ATMEL library, but where is the fun in that?

Step 2

Create a new package by clicking the Package tool and name the package DIP28. Click OK and then Yes to create a new package. The technical drawing for the AVR DIP package is located on page 437 of the ATmega48-328 AVR datasheet (*http://bit.ly/176kUzi*). By referencing this drawing, you can appropriately design the library without touching a single part or measurement tool. This comes in handy when a device is present in a library but doesn't contain the desired package type.

Step 3

Because this is a standard 28-pin DIP package, the pins will be spaced at 2.54mm (Figure 5-5) so set the grid spacing to 2.54 mm by clicking the "Change the grid settings" tool. Select the Pad tool, change the shape to "Shape long," and then change the drill to 0.8 mm. Create a row of 14 pads, starting at the origin and moving to the right. Continue making the second row of pads 7.62 mm above the first row, but this time move from the right to left. Each new pad is given a pad number, and the proper pin number sequence is maintained by placing them in order. You can check the pad number by selecting the Info tool, clicking a pad, and then noting the name.

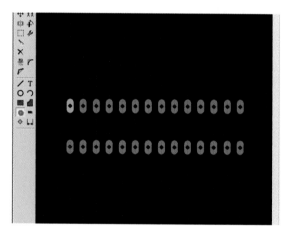

Figure 5-5. *Place the pads in the correct position and verify their location by using the Info tool.*

When designing packages for fine-pitched surface-mount parts and parts without pins, elongate the length of the pad to allow for easier hand soldering. Double the length of the manufacturer's specified pad size and position them so that the pad extends beyond the perimeter of the part.

Step 4

Create the device outline by changing the grid spacing to 0.635 mm and selecting the Wire tool. Change the layer to tPlace, the style to "Right Angle Up Bend," and the line width to 0.127 mm. Start the line at (−1.27, 0) and continue it to (34.29, 6.985). Create another line from (34.29, 6.985) and continue it to (−1.27, 0). Create a circle using the Arc tool starting at (−1.27, 5.080) and completing at (−1.27, 2.540) with a line width of 0.127 mm and a radius of 1.27 mm. This arc will indicate the IC orientation on the silkscreen of the circuit board.

Step 5

Finally, add the name and value placeholders to the package by selecting the Text tool and add ">VALUE" and ">NAME" text strings (Figure 5-6). Change the layer on which these text strings reside by using the Info tool and setting the layer for >VALUE to tValues and >NAME to tNames. These will later auto-complete to the proper name and value when the device is created. This is a good time to save your work.

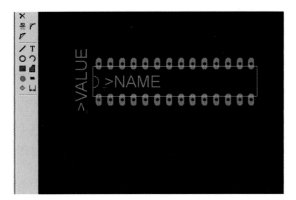

Figure 5-6. *The completed part can be later edited for correctness. Be sure to update the library after editing to reflect the changes in the projects that use the part.*

Create the Symbol
Step 1

Create a new symbol by selecting the Symbol tool and name the symbol ARDUINO328. Click OK and then Yes to create a new symbol. Using the default grid spacing of 0.1 in, select the Pin tool and create a column of 14 pins from (0.0, 1.3) to (0.0, 0.0). Verify that the green circle attached to the pin is facing to the left. This indicates the connection point used when drawing a schematic. Change the orientation of the pin by right-clicking the mouse until the orientation is opposite of the first pins and the green circle points to the right. Make a second row of pins from (1.1, 0.0) to (1.1, 1.3).

Step 2

Change the name of the pins using the "Define the name of an object" tool to the names in Table 5-4. Create an outline around the inside of the pins by using the Wire tool on the Symbols layer. Use the "Draw a text" tool to add >VALUE and >NAME text strings, placing them accordingly (Figure 5-7).

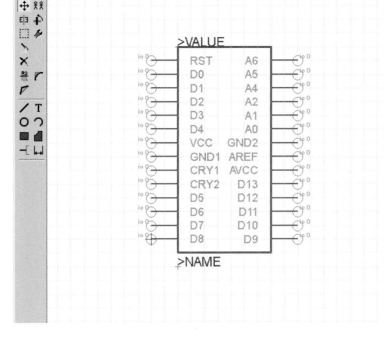

Figure 5-7. *Designing the symbol to reflect the appearance and configuration of the physical part helps you conceptualize the board design prior to layout.*

Table 5-4. Arduino Pin Reference

Pin Column 1	Name	Pin Column 2	Name
1	RST	28	A5
2	D0	27	A4
3	D1	26	A3
4	D2	25	A2
5	D3	24	A1
6	D4	23	A0
7	VCC	22	GND2
8	GND1	21	AREF
9	CRY1	20	AVCC
10	CRY2	19	D13
11	D5	18	D12
12	D6	17	D11
13	D7	16	D10
14	D8	15	D9

Create the Device
Step 1

Use the Edit tool to create a new device named ARDUINO328. In the editor, select the Add tool and add one previously created AR-DUINO328 part. At the bottom right of the screen, click New and add the DIP28 package variant to the drawing. Click Connect and assign the appropriate Pins to Pads by using the "Arduino Pin Reference" table and by clicking connect when the correct pins are highlighted. Click OK when complete.

Step 2

Select the "Define attributes" tool and change the Name to ARDUINO328 and the Value to DIP, and then click OK. These values will show up when the device is imported into the schematic and board layout (Figure 5-8). Save your work, and your library is complete! Remember to enable your library by clicking the little circle next to the library's name in the Control Panel, changing it from a small clear circle to a large green one. Your library is now ready for use.

Schematic Layout with EAGLE

You can use schematic layout software to present the circuit in a way that illustrates each component and its interface with the rest of the components. EAGLE offers a series of tools with which

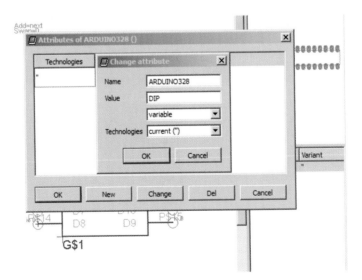

Figure 5-8. Finalize the device by setting attributes.

you can design and manipulate your circuit in preparation for layout.

Schematic Layout Tools

Wire

Each electronic component inserted into the design area has one or more pins. These pins serve as the connection point to other electronic components, connectors, power, and ground. To make these connections, you use the Wire. With this tool, you can select the visual characteristics of the wire, including how it follows the cursor and where it ends. When completed, various other tools, like the Show Objects tool, make it possible for you to highlight any specific trace and follow its path for verification.

Junction

The Junction tool helps to illustrate a divergence in a wire. You can use it to identify if a wire is overlapping or actually connected. This commonly occurs when two or more devices share a connection.

Name

You use the part Name tool to identify parts and wires as they are seen on the circuit board. Devices are commonly identified with a letter prefix followed by a number: R for resistor, C for capacitor, L for inductor, LED for LEDs, IC for ICs, S for switches, H for headers, VCC for positive power, and GND for ground.

Value

You use the part Value tool to assign a numerical value or part number to a device. This is useful when generating a bill of materials because it allows for quick identification of a component's value as well as tallying parts with identical values.

Add

You use the part library in the schematic layout tool to import devices and display a graphic representation of their electrical interface. This library should contain all of the devices necessary to complete the design, and you can use the Add command to include new parts when necessary.

ERC

Every design needs to adhere to predetermined design rules. The Electrical Rule Check tool looks for potential errors and highlights them when found. You can edit this tool to meet the needs of the design and comes preconfigured.

Layer

The Layer function sets the specific design layer on which you are currently editing. The schematic layers include both design and reference artwork and you should check them prior to drawing.

Project: Creating a Schematic in EAGLE

Now that you are familiar with the method in which parts are designed and kept in libraries, it's time to create a schematic (Figure 5-9). This project takes you through the schematic design of a simple breadboardable Arduino.

Step 1

Start by listing all of the features you will want to include in your design, similar to that shown in Table 5-5. This list will help determine the types of components, their quantities, and ultimately, the cost.

Table 5-5. Breadboardable Arduino feature list

Feature	Component
Arduino-based	ATmega328
	16 MHz crystal
	2 × 22 pF Ceramic caps
	10 KOhm Reset pull-up resistor
	Reset switch
	0.1 uF DTR cap
5-VDC	LM2940 LDO regulator
	100 uF 16 V electrolytic cap
	0.1 uF 16 V ceramic cap
External power	2.0 mm barrel jack
	1N4001 rectifier diode
Power indication	3 mm green LED
	330 Ohm resistor
Status indication	3 mm yellow LED
	330 Ohm resistor
Breadboardable headers	24 × 0.1 in male headers
ISP port	3 × 2 × 0.1 in male headers
FTDI interface	6 × 0.1 in right-angle male headers

Step 2

In the Control Panel, create a new schematic by selecting File→New→Schematic and save the file as *Breadboardable Arduino.sch*. To maintain organization and aesthetics, always insert a Frame into the design software by opening the Add tool and selecting A4L-LOC. Insert this frame into your drawing at the origin (Figure 5-10).

Using Table 5-6 as a guide, import the parts into your drawing. The arrangement of the parts is arbitrary at this point.

Table 5-6. Breadboardable Arduino library reference

Component	Library	Name
ATmega328	Arduino	ARDUINO328
16 MHz crystal	special	XTAL/S
22 pF ceramic cap	rcl/C-US	C-US050-024X044
10 KOhm reset resistor	resistor/R-US	R-US_207/7
0.1 uF DTR cap	rcl/C-US	C-US050-024X044
Reset switch	switch-omron	10-XX
LM7805 LDO regulator	linear/78*	7805TV
100 uF cap	rcl/CPOL-US	CPOL-USE2,5-6E
0.1 uF cap	rcl/C-US	C-US050-024X044
2.0 mm barrel jack	con-jack/JACK-PLUG	JACK-PLUG0
1N4001 diode	diode	1N4004
3 mm green LED	led/LED	LED3MM
3 mm yellow LED	led/LED	LED3MM
330 Ohm resistor	resistor/R-US	R-US_207/7
26 × 0.1 in male headers	con-lstb	MA13-1
3 × 2 × 0.1 in male headers	con-lstb	MA03-2
6 × 0.1 in right-angle male headers	con-lstb	MA06-1
VCC	supply2	VCC
Ground	supply2	GND

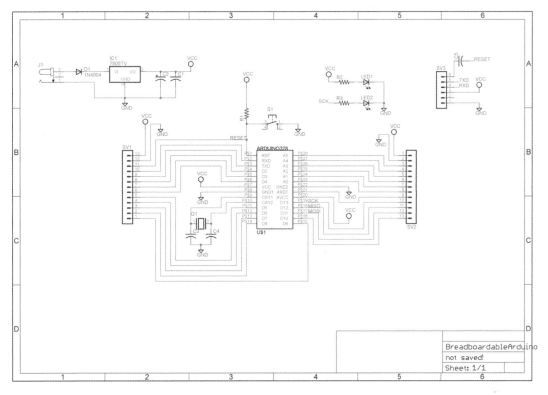

Figure 5-9. *The completed schematic.*

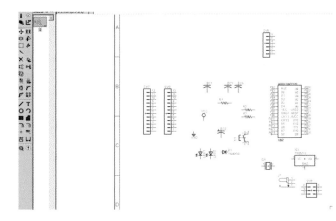

Figure 5-10. *Insert parts into frame and organize into groups with respect to their role within the circuit.*

Table 5-7. Arduino to header connections

Header 1	Arduino	Header 2	Arduino
13	VCC	1	VCC
12	GND	2	GND
11	RST	3	A5
10	D0	4	A4
9	D1	5	A3
8	D2	6	A2
7	D3	7	A1
6	D4	8	A0
5	D5	9	AREF
4	D6	10	D13
3	D7	11	D12
2	D8	12	D11
1	D9	13	D10

Step 4

Position the AVR in the center of the frame and the 12-pin headers on each side. You can rename the headers H1, H2, and so on by using the Name tool or leave them as the default SV1, SV2, and so forth. These headers will act to break the AVR's pins out for proper circuit board spacing. This design will supply VCC and GND to either side of the circuit board, and without the need for breaking out the crystal pins, you need only 24 pins. Using the Wire tool, begin connecting each of Arduino's pins to the corresponding header pin, as specified in Table 5-7. Ensure that these lines do not overlap and are neat and tidy. Complete the connections by attaching the crystal, caps, reset circuit, and power bus indicators.

Step 6

Construct the power circuitry and attach it to the power bus (Figure 5-11). Normally, you can just use the Wire tool to connect these components, but it is a lot cleaner to add short lines to the connections and attach them to the same bus by using the Name tool. Construct the indicators and attach one to the power bus and the other to D13. The last step is to attach the interface circuitry to the Arduino. Save your work and the schematic is complete.

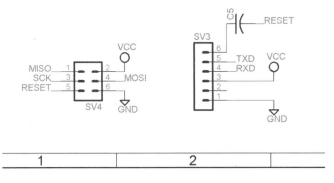

Figure 5-11. *Attach the interfaces by using the Wire tool.*

Board Design with EAGLE

The proper layout of the board is just as important as a functional schematic. After the schematic has been completed and verified it can be exported to the circuit layout tool. For the board to function as intended, you must take both electrical design and mechanical constraints into consideration.

Board Design Tools

Wire

The wires illustrated in the schematic design are meant to establish the circuit's electrical connections. These connections are then imported into the board layout as *air-wires*. These air-wires provide a visual reference for establishing physical traces.

Traces that extend to the opposite side of the circuit board are connected with a plated *via*. The shape and diameter of the via can be selected as an extension of the Trace tool. Vias can be "buried" under solder mask by setting the mask threshold above the diameter of the via in the Design Rules.

Keep trace widths above 8 mil and with 10 mil spacing. These two parameters help prevent both potential mechanical and electronic issues.

Ratsnest

As components are moved around the design area, the air-wires maintain their original reference. You can use the Ratsnest tool to de-clutter the screen and reroute the location of the air-wires, simplifying their path.

Layer

Using the Layer tool, you can select all elements of a single or two-sided board. As Table 5-8 illustrates, these layers include everything from traces to drills.

Table 5-8. Important layers and their function

Layer	Function
Top	Top layer copper
Bottom	Bottom layer copper
Pads	Top/bottom layer pads
Vias	Top/bottom layer vias
Unrouted	Yellow missing air-wire connections
Dimension	Board outline and routes
tPlace	Top component silkscreen
bPlace	Bottom component silkscreen
tOrgins	Top component origins
bOrgins	Bottom component origins
tNames	Top component names
bNames	Bottom component names
tStop	Top solder mask
bStop	Bottom solder mask

Smash

The position of component labels is established within the library file. You can edit these elements on a part-to-part basis by using the Smash tool, which allows for the repositioning of individual part labels.

Auto

EAGLE has the ability to automatically route the traces around the board. This tool serves as a good starting point for beginner circuit board design. When the design is complete, you can add the remaining missing connections by hand, or you can alter the layout to allow for complete autorouting.

Polygon

You use the Polygon tool to draw polygon shapes on any layer. This is especially useful for creating ground pours around the board. You do this by drawing a polygon on the top/bottom layer around the dimensions of the board. Change the name of the polygon to GND and the polygon fills in the spaces between parts and connects itself to every ground connection.

DRC

The design rules are parameters that dictate how components interact with the board layout and what is allowed in the design. When the check is run, the board is analyzed against the design rules and any exceptions are flagged. These rules (Table 5-9) help to determine the manufacturability and proper functionality by limiting certain design characteristics.

Table 5-9. Useful DRC settings

Tab	Setting	Parameter
Clearance	All	10 mil or equivalent
Distance	Copper/dimension	10 mil or equivalent
Sizes	Minimum width	10 mil or equivalent
Masks	Limit	25 mil (buries vias)

Project: Creating a Board Layout in EAGLE

Now that you have created a schematic within the layout tool, the components can be positioned and connected within the board layout tool (Figure 5-12). This tool serves as the last design step prior to physical creation. From here, you can export boards as images for hand etching with the toner-transfer process or send them to the CAM processor to be converted into files necessary for professional manufacturing.

Step 1

In the Schematic Layout tool, create a new board layout from the schematic by clicking on the Board button. This will create a new board layout containing all of the components and their air-wire connections. Change the grid dimensions to inches and size to 0.0125 in by selecting the Grid button. These units can be changed later to whichever system suits your design. Edit the dimensions of the board by selecting the Info tool, changing the dimensions of the top and bottom lines as detailed in Table 5-10.

Step 2

Align the Arduino IC to X:1.55 and Y:0.3 using the Info tool. This positions the Arduino in an optimal location for making all of the necessary interconnects. Position the crystal at the far right of the board so that it's air-wires orient with the Arduino's pins. Position the crystal's capacitors near the top and bottom of the capacitor. Optimally, the crystal and capacitors should be located as close as possible to the crystal pins, but in this case, size limitations trump circuit design optimization. Select the Wire tool and change the wire width to 0.010 in. Route the traces for the crystal and accompanying capacitors. These traces are important and therefore require symmetry and isolation from nearby parts.

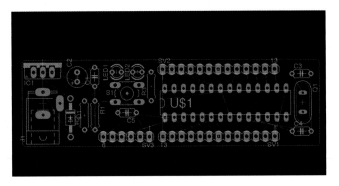

Figure 5-12. *Board layout project.*

Step 3

Position the parts at the locations illustrated in Table 5-11. Orient parts by using the Move tool, right-clicking until the part is rotated to the correct position.

Table 5-10. Breadboardable Arduino top-line dimensions

	X (in)	Y (in)
Top from	3.2	0.9
Top to	0.0	0.9
Bottom from	0.0	0.0
Bottom to	3.2	0.0

Table 5-11. Breadboardable Arduino component positions

Name	X (in)	Y (in)	Angle
IC1	0.225	0.875	0
J1	0.225	0.45	90
D1	0.525	0.3	270
C1	0.7875	0.725	90
C2	0.6	0.725	90
C5	1.1375	0.35	180
H1	2.125	0.01	180
H2	2.125	0.8	0
H3	1.125	0.1	180
R1	0.7875	0.325	270
R2	0.675	0.325	90
R3	1.4	0.625	90
S1	1.1375	0.55	0
LED1	1.025	0.775	180
LED2	1.25	0.775	180

Step 4

Select the Polygon tool and set the width to 0.016 in (Figure 5-13). Create a polygon that completely covers the dimensions of the board. Change the name of the polygon to "GND" and select the Ratsnest tool. This tool will complete the ground pour and recalculate the shortest path for the air-wires.

Figure 5-13. *A solid ground pour on two-layer designs helps to reduce cross-talk between signal lines and requires less etchant if made by hand. Correct breaks and gaps by adjusting the position of components and traces.*

Step 5

Select Wire tool and set the width to 0.024 in. Route the power traces and use vias to indicate when the trace changes sides of the board. Because this board is designed to be single sided, any traces along the top of the board will be made with bare wire jumpers. It is always best to draw these first because their width is larger in order to accommodate greater current loads.

Step 6

Complete the layout as shown in the diagram (Figure 5-14). Ensure that 90-degree trace angles are limited and the spacing between traces is no less than 0.01 in. Use the Ratsnest tool to verify that there are no more air-wires and run the DRC check. This program checks the board for any violations of design rules and flags necessary changes. Correct any errors and the board is complete. You can now print the board for toner transfer or send it to a circuit board.

Soft Circuits

Are you interested in electronics and like to sew? Then soft circuits are for you! Now that we have covered a bit about how circuits are designed, the next step is to construct the project (Figure 5-15). Soft circuits are an exciting way to work with electronics and produce wearable circuits without ever heating up a soldering iron. Rather than relying on conductive circuit board and solder to connect a circuit's components, these circuits rely on conductive thread, conductive fabric, and knots to hold it all together (Table 5-12). These materials are then sewn between a circuit's components, both securing them to the fabric and producing the electrical connection.

Figure 5-15. *A soft circuit project.*

Soft-circuit projects range in complexity from a few simple components, like a battery and an LED, to full-blown microcontrollers commanding lights and sound. Each project begins with the selection of components that not only meet the circuit's needs but are capable of being attached to fabric. With a little modification, you can use most through-hole components, thus opening the door to a world of creative possibilities.

Companies such as Adafruit Industries specialize in wearable electronics and offer a wide range of premade circuits that are project-ready. Find out more at http://bit.ly/12qcAdZ.

Table 5-12. *The soft circuit toolbox*

Tool	Use
Conductive thread	Connecting discrete components
Conductive fabric	Creating large conductive areas
Fabric	Medium for mounting components
Chalk pencil	Outlining the circuit
Wire snips	Cutting the conductive thread
Needle-nose pliers	Bending and wrapping leads
Iron-on backing	Covering conductive components

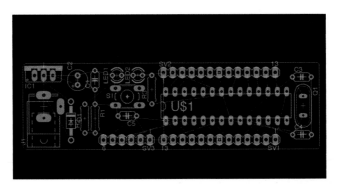

Figure 5-14. *Print out scaled-up copies of your design after you have finished and look for errors.*

Prepairing Your Components

You can modify most through-hole components so that their leads can be sewn onto fabric. You do this by twisting, wrapping, or bending the leads in a way that allows for both mechanical and electronic connections.

Components with long leads can be universally prepared by trimming the leads to around 1/2 inch in length. Use a pair of needle-nose pliers or jeweler's pliers to carefully roll the lead toward the part until only a small amount of straight lead is left (Figure 5-16).

Figure 5-17. Battery holder.

It just so happens that most desktop computers use CR2032 batteries to retain their CMOS settings. These can be easily de-soldered from discarded computers and used as a free battery solution for your next soft-circuit project.

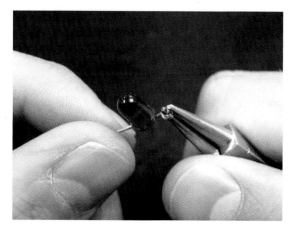

Figure 5-16. *Use a pair of needle-nose pliers to wrap a small length of lead into a coil.*

Battery Holder

The battery is one of the more tricky components to sew onto fabric because they tend to be bulky and heavy. The 3-VDC CR2032 coin-cell battery provides an inexpensive solution to the power problem and provides enough power to illuminate a ton of LEDs, sensors, and other such devices. You can sew these batteries onto fabric by using a common through-hole button holder (Figure 5-17). Simply bend the leads so that they face toward the middle of the button holder and use the two half-hitch knot to secure the conductive thread to the lead. Because these button holders can be a bit wobbly, tack it in place with some nonconductive thread or a small amount of super glue.

Buttons

What's a project without some sort of feedback? Buttons (Figure 5-18) allow for your soft circuit to have an element of control, whether to simply turn the circuit on, or to respond to input. The easiest buttons to sew onto fabric are common tactile *spider* switches. You can sew these by pressing the switch through your fabric and carefully bending the legs 90 degrees to the side. Then, make a couple of stitches around the leg to secure in place with the conductive thread, and you are good to go. Alternatively, you can use the method for connecting ICs because it produces a much more reliable connection.

Figure 5-18. *Soft-circuits button.*

Capacitors

You can sew capacitors (Figure 5-19) onto fabric by trimming their leads to 1/2 in and wrapping them in toward the part so that it lies flat against the fabric. You can hold larger capacitors in place by tacking the body down between the legs with a small amount of nonconductive thread.

Figure 5-19. *Soft-circuit capacitor.*

*Only use sealed capacitors such as ce-
ramic or tantalum rather than electrolytic (the ones that look like soda cans).
Electrolytic capacitors tend to leak their
electrolyte when roughly handled, and
that's not something you want on your
skin.*

Resistors

Prepare each lead by trimming them to 1/2 in and wrap in toward the part by using a pair of needle-nose pliers. Because resistors are so small (Figure 5-20), you can hold them in place by the connections at their leads. Alternatively, you can make a tacking stitch around the body of the resistor.

Figure 5-20. *A soft-circuit resistor.*

LEDs

LEDs (Figure 5-21) can be a bit tricky to sew be-cause they need to be positioned vertically given that they emit light from the top. Prepare an LED by first bending the legs 90 degrees to the side. Trim the lead to 1/2 in and wrap in toward the part using needle-nose pliers. Ensure that the coil sits perpendicular to the body of the LED so that the LED points away from the fabric. Because there isn't a great way for tacking the part in place with thread, you can use a small amount of fabric glue to help keep the LED from falling over.

Figure 5-21. *A soft-circuit LED.*

Transistors

Because transistors have three legs (Figure 5-22), it is important that they are spaced far enough apart to prevent shorting out. Take each outer leg and bend them 90 degrees to the side. Trim the leads to approximately 1/2 in and create loops. Sew the transistor by first tacking the part to the fabric using nonconductive thread on each pin and then run the conductive thread as needed.

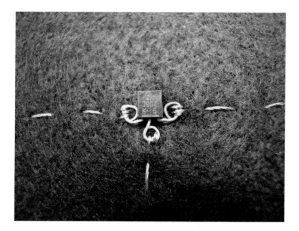

Figure 5-22. *Soft-circuit transistor.*

ICs

ICs are tricky to prepare because their legs are very brittle and close together (Figure 5-23). DIP-style ICs have legs that start off wide at the body of the part and quickly taper to fit into a circuit board. The easiest way to use these devices in your soft circuit involves preparing the legs prior to sewing.

Figure 5-23. *A soft-circuit IC.*

Begin by attaching a length of thread to each leg using the two half-hitch knot and slide a *very small* piece of 1/16 in heat-shrink tubing over the leg and the knot (Figure 5-24). Ensure that the tubing is small enough to allow for the thin part of the leg to pass through the fabric. After each leg is prepared, press the IC into the fabric and bend the thin part of the legs 90 degrees to the side. You can secure large ICs in place with a small amount of hot glue.

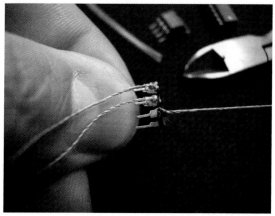

Figure 5-24. *Prepare each leg prior to inserting the IC into the fabric using a two half-hitch knot followed by a small piece of heat-shrink tubing.*

Sewing and Knots

After the leads have been prepared, its time to sew (Figure 5-25). The thread you will be working with comprises many fine strands of stainless steel spun together into thread. The thread is smooth to the touch and behaves sort of like fishing line and is therefore resistant to being tied into a knot. This tends to be a bit of a problem because you are relying on this thread to both electrically connect each of your components and hold them in place. Each step of this process requires a different type of knot, and remember, conductive thread is conductive, so keep your project tidy and free of loose threads. You can use the following knots to get your project started and feel free to experiment with different methods of connecting and securing your components.

Figure 5-25. *Sewing a resistor.*

Starting the Stitch

The double-overhand knot is a simple way for producing a small ball at the end of your string that prevents it from passing through your fabric when pulled tight (Figure 5-26). Make this knot by simply making a loop in the thread and then wrapping the end of the thread a couple of times around the loop. Pull tight to secure the knot and trim the excess thread.

When securing components in place, start with the double-overhand knot and sew around the

component's lead a couple of times to both hold the component in place and produce a good conductive joint.

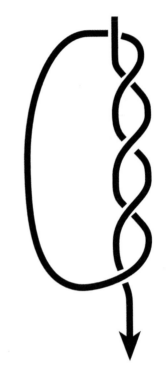

Figure 5-26. *Begin sewing by tying a small double-overhand knot at the end of your thread as shown.*

Sewing

The easiest method for connecting components is to use a running stitch (Figure 5-27). You do this by weaving the threaded needle across the fabric, making relatively long stitches. This is easier than individually creating each stitch and results in a straighter line.

As your designs become more and more complex, you might find a need to overlap stitches (Figure 5-28). Because the conductive thread is, well…conductive, it is imperative that it does not touch adjoining components and thread. To create an overlap, make the initial stitch between the respective components, and when the second stitch is close to the initial stitch, create a stitch that jumps to the opposite side of the fabric. This puts a small amount of fabric between the two wires and helps to isolate them.

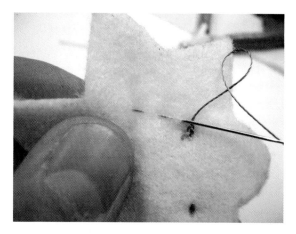

Figure 5-27. *You can make connections between components by using a quick running stitch.*

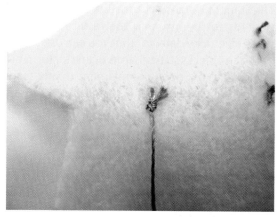

Figure 5-29. *Secure components with legs by using a simple two half-hitch knot.*

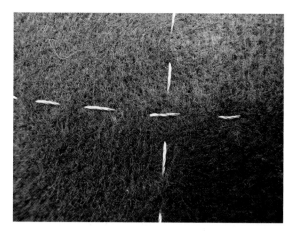

Figure 5-28. *Overlapping stitches without letting them touch allows for you to create much more complicated designs without the risk of shorting out the connection.*

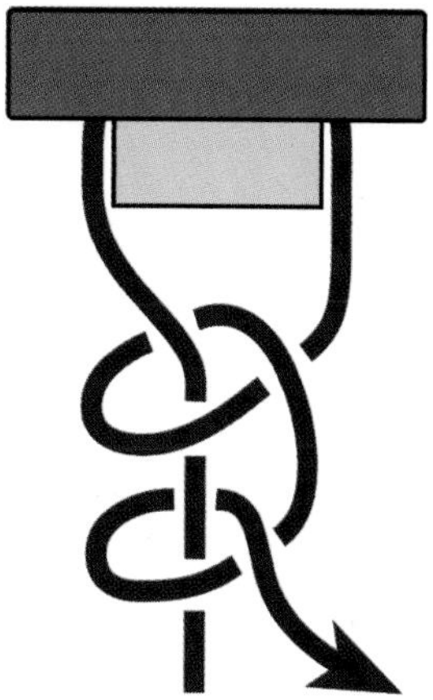

Figure 5-30. *Make your knot by first wrapping the thread around the bent lead and making two half-hitches.*

Connecting Legs and Finishing Stitches

Some components don't have leads that can be wrapped into loops, and it is necessary to tie the thread directly to the pin (Figure 5-29). Simply use a two half-hitch knot (Figure 5-30) to secure the thread to the leg after it has been bent over. This not only makes a good electrical connection, it secures the component in place, as well.

Try adding a drop of super glue to the knot if you find that they loosen over time. This also stabilizes the joint when flexed and helps to keep out corrosion-forming moisture.

Project: The Static Sticky

Sticky's are simple circuits that are made up of only an LED and a coin-cell battery (Figure 5-31). When the battery is connected to the LED, the LED illuminates, and you can stick it just about anywhere, making for some mischievous fun.

The Static Sticky project adds just two more components to the mix, and the result is a handy static electricity detector that you can use to visualize an object's static charge, or just for fun. This device is also a good complement to any electronics bench because it helps notify you when unwanted static charges are present around your sensitive electronics.

Figure 5-31. *The Static Sticky.*

Materials

Project Materials List		
Material	**Qty**	**Source**
Felt	1	Arts and crafts store
1 MOhm resistor	1	Electronics supply store
MPF-102 FET	1	Electronics supply store
CR2032 battery holder	1	Electronics supply store
Red LED	1	Electronics supply store

Makerspace Tools and Equipment
Conductive thread
Needle-nose pliers
Pencil
Scissors
Sewing needle
Wire snips

Procedure

Step 1

Begin by preparing your component's leads as previously described and place them on top of your felt, as shown in Figure 5-32. When they are positioned appropriately, draw a star pattern around them with your pencil. Feel free to be creative here, how about a moose or maybe an amorphous blob? After your outline is complete, draw a faint outline around your parts so that you can easily reposition them.

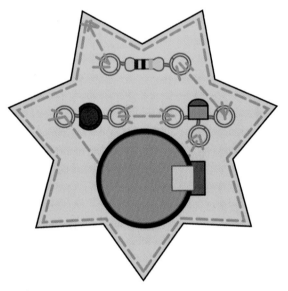

Figure 5-32. *Arrange the components as shown, and make sure you have enough room to sew.*

Step 2

Cut out the pattern using your scissors and attach the battery holder (Figure 5-33). Depending on the type of holder you are using, you might need to secure it better in place by applying a small amount of hot glue. Continue to connect the remaining components and finally sew a perimeter of conductive thread around the entire shape. This acts as the "antenna," giving your sensor better range.

Figure 5-33. *Draw an outline around your parts to ensure you have enough room.*

Step 3

Connect the battery and test the circuit. If everything is properly connected, the LED should gradually illuminate as it draws closer to a static charge. Pretty cool, huh?

Making Circuit Boards

Before you can make any circuit board, it has to have a solid schematic and a board design that illustrates the size of the board, position of components, and layout of the traces. These features ultimately dictate how the board will be manufactured, and you can tailor them to the specific type of manufacturing. For example, if the circuit board, commonly referred to as PCB (for printed circuit board), is going to be made by hand using the toner-transfer process, it is a lot easier to keep the board single-sided and to increase the size of the pads and traces.

Generating CAM Files Within EAGLE

For a circuit board manufacturer to create your boards, it needs a series of files that represent each layer of your project. Exporting these files from EAGLE is a simple process and involves the use of the CAM processor. These CAM files are then sent to a manufacturer for verification and will often generate a quotation regarding unit cost (and cost breaks, based on volume).

Step 1

You need to run two jobs from within the CAM processor in order to produce the proper files for manufacturing. The first, *excellon.cam*, generates the drill information. Open the job file and click Process Job to generate the files. The second, *gerb274x.cam*, generates the remaining files. Open the job file and verify the following configuration presented in Table 5-13.

Table 5-13. EAGLE 274X configuration

Page	File Extension	Layers	Function
Component side	*.cmp	Top, Pads, Vias	Top copper
Solder side	*.sol	Bottom, Pads, Vias	Bottom copper
Silk screen CMP	*.plc	Dimension, tPlace, tNames	Top names
Solder stop mask CMP	*.stc	tStop	Top solder mask
Bottom stop mask CMP	*.sts	bStop	Bottom solder mask
Board outline	*.outline	Dimension	Board outline

Step 2

Create any missing files and double check that none are mirrored. When complete, click Process Job and the CAM processor will produce the remaining boards. These files should be compressed in a ZIP file for proper upload to the circuit board manufacturer.

Ladyada has a great circuit board manufacturer database that displays each manufacturer's cost/in² in addition to information about each, such as manufacturers and cost.

Professional Manufacturing

Most circuit boards are made in a method that involves laminating sheets of nonconductive material between thin sheets of copper foil or other conductive materials, such as nickel and aluminum, that is coated with a photosensitive emulsion. Standard circuit board thickness is 0.063 in (1.57 mm). You can adjust this depending on application.

This copper-clad panel is then placed in a CNC drilling machine that drills all of the holes and vias that will be used for mounting components and routing traces. The holes are deburred and the walls of the vias are roughed so that they can be plated. The panel is cleaned and the circuit image negative is exposed onto the surface. This step exposes the traces to an intense ultraviolet light that polymerizes the emulsion so that it can be easily washed away. After the exposed emulsion is removed, the panel is submerged in an etchant solution that removes the exposed copper from the board. The panel now has a complete circuit design on its surface, and if you have more than one copper layer, it's ready for *lay up*. This process involves aligning the layers in a heated press and compressing them until they adhere.

Plating occurs when the panel is connected to a long conveyor system that submerges it in a tank that electroplates any exposed conductive material. The plated material consists of mainly nickel or gold, which helps to protect the copper from oxidizing and can improve conductivity. Alternatively the panel can be coated in flux and immersed in the plating solution. Table 5-14 presents some typical surface finishes and their characteristics.

Table 5-14. Common surface finishes

Material	Notes
Bare copper	Low cost but prone to corrosion
Immersion tin	Standard coating with good wettability
Immersion gold	Best coating with optimal wettability

The panel is then coated with solder mask and silkscreened with any part names and build information. This coating helps to prevent electrical shorts and protects the unexposed traces from oxidation. The final step in the process is to route out any large sections of the board and prepare the boards for separation. This consists of either routing out most of the perimeter and leaving a few perforated holes for support, routing out the boards completely, or scoring the panel for easy separation.

Ladyada provides a handy calculator for determining circuit board manufacturing cost based on dimensions and quantity, found here: http://www.ladyada.net/library/pcb/costcalc.html.

Project: Toner-Transfer Circuit Board

The method for producing the files required for the toner-transfer process (Figure 5-34) is completely different than that for manufacturing. For the toner-transfer process, only the top and bottom copper layers are used, or only the bottom copper in the case of a single-sided board. This process involves transferring a full-scale copy of the copper layer onto a bare copper-clad circuit board and submerging the board in an etchant. The etchant, typically ferric chloride, dissolves away any exposed copper, leaving the desired traces behind. Toner-transfer circuit board fabrication is a great way to produce low-quantity prototype boards, and with enough practice, even fine-pitch surface-mount designs are possible (Table 5-15).

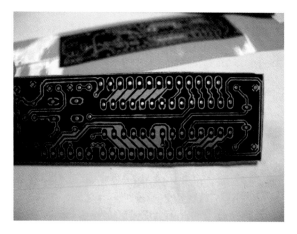

Figure 5-34. *The toner-transfer project.*

Table 5-15. Toner-transfer paper types

Paper Type	Source
Kapton on printer paper	Homebrew
Press'n'peel blue	http://www.techniks.com/
Glossy magazine paper	The mailbox
Glossy laser photo paper	Office supply store

This project requires that safety glasses be worn throughout its entirety.

Materials

You can find the files for this project at Thingiverse.com (*http://bit.ly/1dtK67G*).

Project Materials List		
Material	**Qty**	**Source**
Bare copper board	1	Electronics supply store
Toner-transfer paper	1	See Table 5-15

Makerspace Tools and Equipment
Ferric chloride
Gloves
Heat press
Laser printer
Plastic tub or sealable sandwich bags
Safety glasses

Procedure
Step 1

Start exporting a MIRRORED 600 DPI image of each copper layer onto transfer paper using a laser printer (Figure 5-35). The transfer paper Table 5-15 allows for the toner to transfer to the copper when pressed and heated. Cut out an appropriately sized piece of one or two-sided bare copper circuit board and clean the surface by using rubbing alcohol to remove any dirt or oil that might inhibit the transfer. Adjust the heat press to 300°F (150°C) and secure the artwork to the board using a small amount of masking tape. This tape tends to leave sticky residue on the

copper after the heating process, so the less you use, the easier it is to clean.

Figure 5-35. *Make sure your resolution is 600 DPI or greater and the quality is set to maximum. This will help deposit the most amount of toner possible.*

Step 2

When the press has finished heating, position the board in the center of the bed, clamp the press, and let it cook for 10 minutes. Carefully remove the board from the press and allow it to cool on a heat-resistant surface, making sure not to disturb the transfer paper. After the board cools, very carefully remove the tape and transfer the board into a small tub of water for about 10 more minutes. When you detect signs that the water is absorbing into the paper, gently peel away the transfer paper until all of the toner is left exposed. There is a lot of variation in how well the toner sticks to the copper, so this process might take a few tries to get a clean transfer.

Breaks and gaps in the toner traces can be fixed by using a permanent marker (Figure 5-36). Gently fill in the gaps with a thick layer of ink and let dry before etching.

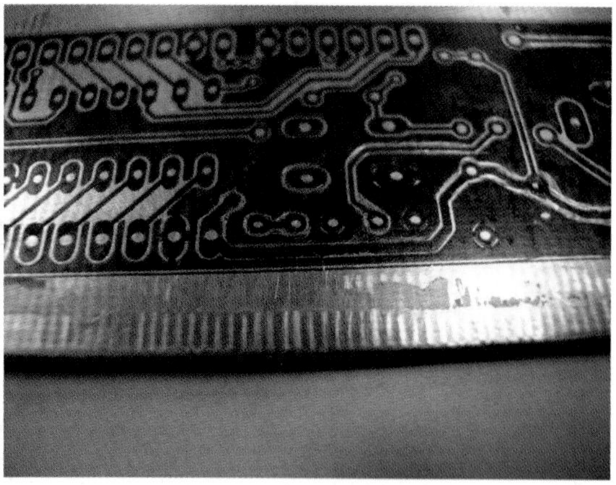

Figure 5-36. *Repairing the toner traces.*

Step 3

Prepare a bath of etchant and submerge the board in the solution. This solution needs constant gentle agitation to deliver the best results. Alternatively, you can place the board inside two sealable sandwich bags. Just add etchant and the boards, remove the air, and rub gently until complete. Keep a close eye on the etching process because there is a short window between etching just enough and over etching the board.

After the etching process is complete return it to the bottle for reuse or properly dispose of the used etchant by contacting your local hazardous waste disposal company. *Do not pour it down the drain because the dissolved copper is harmful to the environment!* Rinse the board with soap and water and inspect the board for errors. These can be corrected by using a soldering iron and enough solder to bridge the gap. If the gap is too big, you can solder the leg of a resistor in line with the trace.

Step 4

Finally, drill the holes by using a drill press and an appropriately sized bit (Figure 5-37). If you were making a double-sided board, you can make the vias by using either rivets or carefully soldering a small piece of wire to

either side. Visually inspect the board again, and if all checks out, it's time to populate it!

Figure 5-37. *Populate your board and check for short-circuits with a multimeter. If everything checks out, power it up and see if it works.*

Working with Microcontrollers

From your calculator to the most complex robot, microcontrollers are responsible for providing a hardware interface to executable code. You can think of a microcontroller as a tiny computer. It has a processor, storage, RAM, input and output interfaces, and requires power. Individually, these components might be exponentially less capable than those found in your everyday computer, but their function is identical.

Microcontroller Fundamentals

Microcontrollers have been at the forefront of Makerspace technology since the beginning. Their ability to bridge the gap between the virtual world run by software and the physical world run by hardware makes them the backbone of most Maker projects. By understanding the microcontroller's feature capability, your Makerspace will be better prepared to integrate them into future projects. All microcontrollers contain a series of features that allow for full system operation with the need for very few external components.

Processor

The processor is responsible for completing the calculations required for executing software and controlling the physical hardware interfaces. The speed at which the processor completes these calculations is controlled by a frequency generator in the form of a tuned-crystal or ceramic oscillator. Many people choose to run their microcontroller at the fastest speed possible. Faster is better! Right? Well, not necessarily. When a microcontroller operates at a higher clock speed, it requires more power. This might not be a problem if the microcontroller is powered from a wall outlet or USB. But, if the project is to be battery powered, energy consumption is everything and a lower clock speed might be the better choice.

Memory

Computers today can store many terabytes of data and have many gigabytes of RAM. That is a *lot* of storage, and the more you become familiar with memory on a microcontroller, the more you will begin to appreciate its sparseness. The quantity of memory on a microcontroller varies greatly between make and model and is one of the primary variables that dictates the different part numbers for the same general microcontroller. Table 5-16 illustrates the available memory on the popular ATmega328 microcontroller from ATMEL's AVR line.

Memory that is labeled as volatile requires constant power to retain its information. This memory type is often used in situations that do not require long-term memory retention. The advantage of RAM is its ability to quickly perform an almost unlimited read/write cycles.

Memory labeled as nonvolatile retains its information even after power has been removed. This memory type is used in situations that require long-term memory storage. In contrast to the virtually unlimited read/write cycles found with RAM, the AT-

MEGA328 is designed to withstand only 10,000 flash and 100,000 EEPROM read/write cycles.

Table 5-16. ATmega328 memory features

Type	Quantity	Purpose
Flash Memory	32 KB	Nonvolatile memory that contains the executable program and can often be used to store variable values
RAM	2 KB	Volatile memory used for the storage of variable values
EEPROM	1 KB	Nonvolatile storage for general purpose use

I/O

A microcontroller's input and output, or I/O, capability provides the interface with the outside world. These interfaces can detect things like button presses, heartbeats, light intensity, and can control things like LCDs, lights, relays, and motors. More sophisticated protocols such as I^2C and SPI allow for high-speed interfacing with a vast array of ICs, and you can even use them for networking multiple microcontrollers.

ATMEGA328 I/O Capability		
Protocol	**Ports**	**Features**
Digital I/O	23	40 mA max
SPI	2	Speed relative to clock
TWI	1	100 or 400 kilobit/second (kbps)
1-Wire	n	Software defined
UART	1	2400–115200 baud
ADC	8	10-bit resolution
PWM	6	8- and 16-bit resolution

Power

The typical microcontroller can operate between 1.8 V and 5.5 VDC, depending on type. The actual operating voltage is dictated both by the microcontroller as well as the hardware with which it interfaces. In most situations, the power applied to the microcontroller is regulated through either a resistive or switching power supply. This makes it possible for the microcontroller to operate with a supply voltage greater than its operating voltage, as demonstrated by powering an Arduino with a 9 V battery. Resistive or linear regulators, like the popular LM7805, change their resistance internally to regulate the voltage. This method uses Ohm's law to determine the output voltage based on current draw and adjusts the internal resistance accordingly. The resulting "left over" energy is burned off as heat, which is why this type of regulator gets so hot under large current loads. Switching regulators modulate their output voltage by implementing a frequency generator that switches the output voltage on and off thousands of times per second. This method of voltage regulation is significantly more efficient than the resistive type and can even achieve efficiencies of over 95 percent. There are also cases in which you might not need a voltage regulator. It just so happens that you can happily power a microcontroller with two AA-type batteries. Connecting two 1.5 V batteries in series will not exceed most microcontroller's maximum operating voltage and can result in longer system operation.

Analog and Digital Input

Input on a microcontroller is sensed as either an analog or digital signal. Analog signals entering the microcontroller consist of a varying voltage that is interpreted by an analog-to-digital converter (ADC). Digital signals are read as a 1 or 0, indicating a high or low voltage change at the pin. Both of these interfaces are conduits for the microcontroller to connect to and sense the world around us.

Analog

The microcontroller's ADC separates an analog signal into small numerical chunks depending on the interface's bit resolution. For example, an 8-bit ADC with a 5 V reference will separate 5 VDC into 256 chunks. 0 representing 0 VDC and 255 representing 5 VDC. As the bit resolution increases, the more the ADC can sense the change. Typically on-

board ADCs are either 8 or 10 bits and often do not have a separate ADC voltage reference. The ADC voltage reference provides the microcontroller with the voltage range for the conversion. In the case of the ATmega328, the ADC voltage reference can be set to an internal reference of 1.1 V or can be pulled as high as the microcontrollers operating voltage.

Digital

Digital signals are read by using a voltage threshold of around 1.4 VDC. When the input voltage exceeds the threshold, the microcontroller reads a 1, and when it falls below, it reads 0. You use this interface for both slow-switching input such as buttons and switches, or as data input which can switch many thousands of times per second.

Communications

Our ability to communicate with both spoken and written languages has set us apart from the rest of the species on earth. These forms of communication translate directly into the world of microcontrollers and their dedicated communication hardware and software libraries. For a microcontroller to send or receive data in any form, there needs to be an established communication method. This can be as simple as waiting for a high or low signal over a digital I/O line or as decoding parallel lines of encrypted data. Regardless of the communication type, the task is the same. To transfer information between two or more devices.

Information sent as a serial string of binary code serves as the basis for virtually all communication methods. Communication starts with the rise and fall of voltage on a signal line and can range in length from one command bit to the interpretation of billions of bits per second. Typically, the bit stream is sent in the form of 8-bit packets containing a series of 1s and 0s. Each position along the byte represents a base-two multiplier. Counting from right to left, the 8-bit packet can represent any whole number from 0 to 255.

Serial Communication

Serial communication (Table 5-17) is the simplest method establishing communication with your microcontroller. In its most basic form, serial communication requires only three connections: TxD, RxD, and Ground (Figure 5-38). The terminal end of the serial connection, commonly a computer or microcontroller, is referred to as Data Terminal Equipment (DTE) and the client is referred to as the Data Communications Equipment (DCE). It is important to know the relationship between devices because it dictates the connection of the TxD and RxD signal lines.

For stable communication to take place, you need to configure each device to the same data transfer rate, or *baud rate*. Common baud rates in increasing order are 2,400, 4,800, 9,600, 19,200, 38,400, 57,600, 115,200. As data transfer increases in speed so does the chance for communication errors. To tackle this problem the RS232 interface has integrated hardware flow-control lines. Flow control utilizes the DSR, RTS, and CTS data lines to indicate to each piece of hardware when data is ready to be sent and received.

There is a common problem that occurs when interfacing low-voltage microcontrollers with devices operating at RS232 levels (Table 5-18). The RS232 standard relies on voltages between −15 VDC and 15 VDC to represent binary 1 and 0. Microcontrollers typically operate between 1.8 VDC and 5 VDC and will be physically damaged if connected to higher voltages. To achieve safe communication between the microcontroller and a true RS232 device, you need a voltage conditioning circuit.

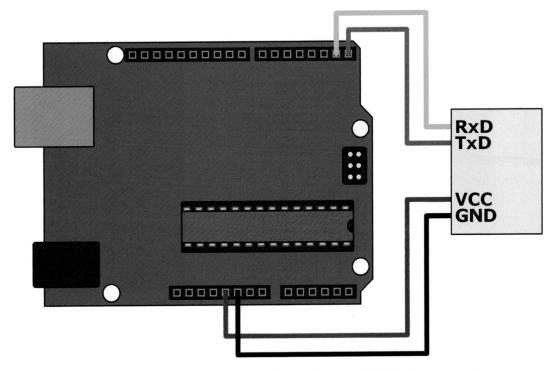

Figure 5-38. *Serial communication requires three connections: TxD, RxD, and GND. The fourth connection, power, is only needed if the client device requires power.*

Table 5-17. RS232 DB9 data and control line assignment

Pin Number	Name	Function
1	DCD	Data carrier detect.
2	RxD	Data Receive.
3	TxD	Data Transmit.
4	DTR	Data terminal Ready. Raised when DTE is powered on.
5	GND	Signal ground.
6	DSR	Data Set Ready. Raised when DCE is ready.
7	RTS	Request to Send. Raised by DTE when data is ready to send.
8	CTS	Clear to Send. Raised by DCE when data is ready to receive.
9	RI	Ring indicator.

Table 5-18. RS232 voltage specification

Tx Signal	Value	Voltage
	Binary 0	+5 to +15 VDC
	Binary 1	−5 to −15 VDC
Rx Signal	**Value**	**Voltage**
	Binary 0	+3 to +13 VDC
	Binary 1	−3 to −13 VDC

The easiest method for connecting an RS232 device to your microcontroller is with a simple voltage divider (Figure 5-39). Although this is not the safest or most stable method, it works well for lower baud rates. The values of the divider's resistors can be calculated by using the following formula to solve for R_1. $R_1 = (V_{in} - V_{out}) \times R_2 / V_{out}$. Calculating for a 5 VDC to 3 VDC divider would yield a 1.8 kOhm for R1 and a 3.3 kOhm for R2. If higher baud rates are needed, you can use a level shifter IC like the MAX 232, which can be used to buck voltages down to the microcontroller and boost them up to the RS232 device.

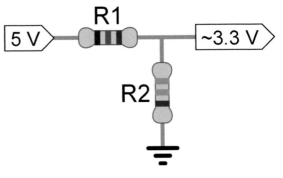

Figure 5-39. The voltage divider.

If you are reading multiple bytes using a loop, make sure you account for the speed in which the bytes enter the buffer. If the microcontroller reads the buffer too fast, it might pick up erroneous bits, or if it reads too slowly, bytes can be lost. You can solve this problem by adding a short delay after each byte is read.

1-Wire

1-Wire serial protocol makes it possible for multiple devices to communicate over a single data line (it is a registered trademark of Maxim/Dallas Semiconductor [Figure 5-40, Table 5-19]). Each network consists of a master and one or more slave devices that typically include battery monitors, thermal sensors, authentication, and memory.

Table 5-19. 1-Wire data and control line assignment

Name	Function
DQ/Data/OWIO	1-Wire data signal line

Device Identification

Each 1-Wire device is factory programmed with a unique 64-bit identification number. This number consists of a 1-byte family code that you can use to identify the device type and function. The device in the above example is a DS2438. You can identify it as such from the 0x26 family code (Table 5-20).

Table 5-20. Example 1-Wire identification number

7	6	5	4	3	2	1	0
0x26	0x5A	0xD5	0x07	0x01	0x00	0x00	0x21

Voltage Levels

1-Wire devices are designed to operate over a voltage range of 2.8 VDC to 5.25 VDC, which makes for convenient use with standard 3.3 VDC and 5 VDC devices. Each 1-Wire bus is required to have an external pull-up to the power supply that stabilizes the logic levels. Some 1-Wire devices can even operate without a direct connection to the power supply. The device gets its power parasitically through the data bus and provides for a circuit that requires only two wires, data and ground.

Communication

Communication begins when the master device sends a reset signal to the bus by pulling it low for a defined period. This pulse alerts the slaves to send a presence signal in response. The master then selects a device by transmitting its unique identification number. Communication can also begin if only one slave is on the bus by issuing a skip-rom command. The device select command isolates the individual device on the bus until the bus is reset again. Now, depending on the device type, the master and slave communicate by passing commands and responses over the bus.

I²C

I²C, or two-wire bus protocol, is a simple, two-wire communications interface that allows for 8-bit, serial, bidirectional data transfer (Figure 5-41). Developed by Philips Semiconductors (now NXP Semiconductor), I²C bus allows for communications speeds from 100 kbps in standard-mode to up to 5 megabits per second (Mbps) in Ultra-Fast mode over only two bus lines, SDA and SCL.

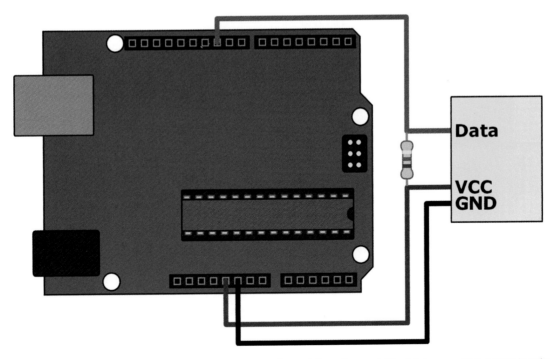

Figure 5-40. 1-Wire communication requires two connections: data and ground. The third connection, power, is only needed if the client device requires power.

I²C Data and Control Line Assignment	
Name	**Function**
SDA	Serial data line
SCL	Serial clock signal

Device Identification

Each device on the I²C bus is software addressable with a 7- or 10-bit address and can assume the role of master or slave. Having multiple masters on the bus allows for more than one device to initiate communication and therefore can be a master-transmitter or master-receiver. Because having multiple masters on one bus can cause collisions if two or more masters initiate the bus, I²C includes collision detection techniques to prevent data corruption.

Voltage Levels

I²C devices are designed to operate over a voltage range of 3 VDC to ~5 VDC and is dependent on the device itself. The communications bus lines must be pulled up to the specified levels for stable communication. 4.7 kOhm resistors are commonly used for this purpose.

Communication

Communication on the I²C bus begins when a master issues a START condition by making a HIGH to LOW transition on the SDA data bus while holding the SCL clock bus HIGH. The bus is now considered to be busy until a STOP condition is set by the master by making a LOW to HIGH transition on the SDA data bus while holding the SCL bus HIGH.

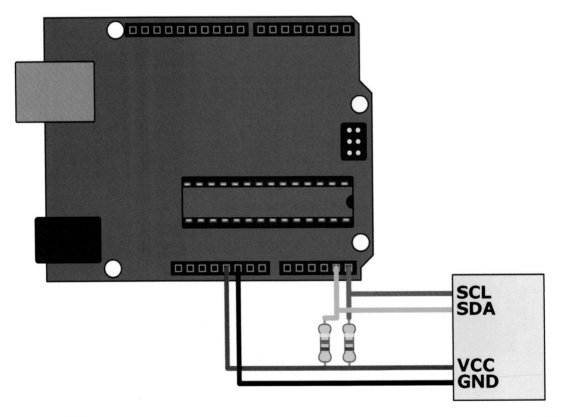

Figure 5-41. *I²C communication requires three connections: SDA, SCL, and GND. The fourth connection, power, is only needed if the client device requires power.*

After the START condition is set, the master then selects the slave device by sending the device's unique address and waits for an acknowledgment, or ACK, bit. If the acknowledgment is received, communication ensues until the master issues a STOP condition. If it is not, the master reads a non-acknowledgment, or NACK, bit indicating a failed communication attempt.

Serial Peripheral Interface

The Serial Peripheral Interface (SPI) is a single-ended serial communications bus designed for short-distance communication between integrated circuits (Figure 5-42). The bus network consists of a master device that controls and communicates with one or more slaves through the use of individual chip select inputs (Table 5-21). Because the speed of data transfer is regulated by the clock signal generated by the master, speeds in excess of 20 Mbps can be achieved.

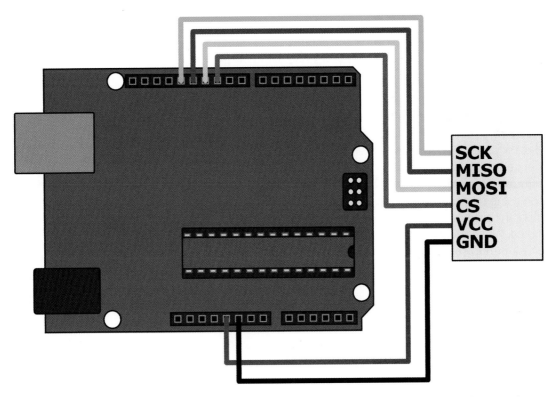

Figure 5-42. *SPI communication requires five connections: MOSI, MISO, SCK, CS, and GND. The sixth connection, power, is only needed if the client device requires power.*

Table 5-21. SPI data and control line assignment

Name	Function
MOSI	Data line from master to slave
MISO	Data line from slave to master
CLK	Clock signal
CS	Chip select control line

Communication

SPI communication is initialized when the master configures the clock by setting the frequency. This frequency needs to be less than or equal to the maximum frequency allowed by the slave. The master then pulls the respective slave's CS pin low. This initializes the slave device, and the master should be configured to wait the designated amount of time required for the slave to generate data or complete the designated task. When the wait period is over, the master issues clock cycles that correlate to data either being transmitted, received, or both simultaneously over the MISO and MOSI signal lines. Data is usually transferred in 8-bit chunks and organized MSB to LSB.

The data transferred over SPI is interpreted based on one of four different SPI modes. Each mode illustrates which edge of the clock signal data is sampled and set with the clock polarity (CPOL) and clock phase (CHPA) registers. SPI can utilize either dedicated hardware or software means for communication, also known as *bit-banging*. When using dedicated hardware, the microcontroller can continue with tasks while data is being handled. While the processor gets dedicated to the task during software communication. Communication completes when the master pulls the CS pin HIGH and the slave enters its standby state.

Interfacing with Hardware

Microcontrollers are capable of both receiving and transmitting information in the form of voltage and data. These ports make it possible for the microcontroller to interface with just about anything, from LEDs to the Internet. By understanding and utilizing these interfaces, the projects in the Makerspace will be able to sense and control even the largest of challenges.

Motor Types

There are a variety of methods for interfacing a microcontroller to a motor. The ultimate goal in this task is to provide the most control with the most amount of protection to the microcontroller's I/O. Because motors often produce large inductive loads, just the act of turning a motor off can burn out a microcontroller without proper protection. Electric motors come in a variety of forms, ranging from pure DC to digitally controlled servos.

Brushed

Brushed DC motors are arguably the most common and widely available (Figure 5-43). These devices operate by spinning a coil of wire (called the armature) within a magnetic field. These can create a lot of electrical noise, and unless you use an encoder, cannot be positioned. They can be driven in a variety of ways including Transistor, Relay/SSR, Darlington Array, and H-Bridge. Depending on how much torque you require to move your project, you can also find some with an attached gearbox.

Figure 5-43. *Brushed DC motor.*

Brushless

Brushless DC motors are simpler in design than their brushed counterparts but require significantly more complex control systems (Figure 5-44). When in operation, brushless DC motors do not produce the same amount of electrical noise as brushed motors. This is because the magnet spins rather than the coil. These motors can be driven by using a transistor array that sequentially cycles the coils. An advantage of this process is that the motor's RPM can be easily calculated based on your pulse frequency.

Figure 5-44. Brushless DC motor.

Stepper

Stepper motors come in two primary configurations: unipolar and bipolar (Figure 5-45). Unipolar stepper motors contain a series of coils that spin a permanent magnet. Each coil is attached to a common *tap* that is typically attached to the positive side of your power source. These typically have five or six wires and, in the case of six-wire motors, the two tap wires can both be attached to the same power source. An advantage of the unipolar motor is that it can be driven by a transistor array, unipolar stepper driver, or bipolar stepper driver, and have the ability to be driven to a specific angle, which provides accurate positioning. Bipolar stepper motors are very similar to unipolar

steppers, although their coils do not have a common tap. This type needs to be driven by a stepper driver or H-bridge as the polarity within the coil switches.

Figure 5-45. Stepper motors.

Servo

Servos contain not only a motor but a torque generating gearbox and sophisticated control circuitry, as well (Figure 5-46). These devices operate by referencing the position of the output to the position of an internal potentiometer. This affords very accurate positioning via the control frequency.

Figure 5-46. Servo motor.

You can modify standard servos for continuous rotation, making them an ideal drive mechanism for robotics. This is accomplished by removing the plastic stop inside the gearbox and decoupling the potentiometer. This makes bidirectional rotation possible as well as speed control.

Driving Motors

Controlling the speed of a motor is a very simple process with Arduino, thanks to a built-in function and technique called *pulse-width modulation*, or PWM. Arduinos have the ability to independently vary the speed of up to six DC motors (on pins 3, 5, 6, 9, 10, and 11) by pulsing a motor on and off using only a few power handling components. PWM is controlled with the *analogWrite(pin, value)* command, which switches the output at a 0 to 255 duty cycle at 490 Hz (this is why motors that are being PWM'd make noise at low duty cycles). The lower the number, the slower the motor's rotation, and vice versa. One of the nice things about the PWM command is that it actually runs in the background of your sketch. When you execute the command, it will continue at that duty cycle until you change it.

Transistor

You can use transistors such as the TIP120 to drive motors and only require a base resistor and a suppression diode (Figure 5-47). These transistors can be controlled by driving a control line HIGH in the case of an NPN and LOW in the case of a PNP. The size of the base resistor depends on the quantity of current allowed to flow through the transistor and is calculated as follows:

$$Rb = Vs / (Imax/hFE)$$

Rb	The base resistor value. This number is often rounded to the closest available resistor value.
Vs	Source voltage.
Imax	Max current through circuit.
hFE	The transistor's forward current gain.

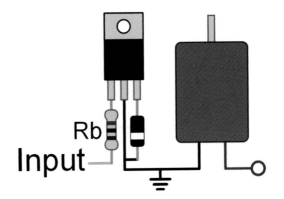

Figure 5-47. *Transistor-based motor controller.*

Darlington Array

Darlington arrays, like the ULN2803, consist of a series of Darlington transistors attached to a common supply. You can switch each transistor individually and some arrays even include internal suppression diodes further reducing the required component count. To drive a motor with a Darlington array, connect the motor's positive terminal to the voltage supply and place the Darlington in series with the motor's ground. The motor can then be switched on and off by changing the logic level at the control pin.

H-Bridge

H-bridges such as the L293D make clockwise and counterclockwise rotation of a DC motor possible. These devices consist of a series of Darlington transistors that can turn on and off the IC's outputs, as well switching their polarity, providing more robust motor control.

H-bridges that do not contain internal suppression diodes should be included to prevent unintended damage to the IC and control circuitry.

You can even use H-bridges to drive bipolar stepper motors, offering a cheap alternative to sophisticated stepper drivers.

The Arduino Development Platform

For novices and veterans alike, the Arduino (Figure 5-48) is a marvelous platform for development. It has quickly become the backbone for many of today's open source projects with its easy to use Integrated Development Environment (IDE) and wide range of feature-expanding "shields." Whether it is the middle school classrooms teaching students to flash LEDs or integration with commercial products, it's not hard to see why Arduino has become so popular. Although there are many other development alternatives, very few, if any, can compete with the wealth of user support, documentation, learning materials, expansions, designs, and variants found with Arduino—not to mention that most of which can be found as open source.

The heart of the Arduino platform contains one of ATMEL's 8-bit ATMega microcontrollers (32-bit in the case of the Due). These microcontrollers provide all of the feature capability the Arduino can deliver, made available through a series user-accessible headers on a wide range of Arduino development board variants.

Figure 5-48. *Arduinos.*

Prototype Arduino

You can create a simple Arduino by using only a few components and a breadboard (Figure 5-49). The parts list in Table 5-22 and the following diagram can be used to make a breadboard Arduino variant that works great for those just starting with Arduino.

Table 5-22. *Breadboard Arduino parts list*

Part
0.1 uF capacitor (located next to the reset switch)
10 kOhm resistor
16 Mhz crystal
22 pF capacitor (located next to the crystal)
ATmega328
FTDI breakout
SPST switch

Arduino I/O Comparison Chart								
Board	**Processor**	**Digital I/O**	**Analog**	**PWM**	**Flash**	**SRAM**	**EEPROM**	**Clock**
Uno/Pro	ATmega328	14	6	6	32 KB	2 KB	1 KB	16 MHz
Leonardo	ATmega32u4	20	12	7	32 KB	2.5 KB	1 KB	16 MHz
Mega	ATmega2560	54	16	15	256 KB	8 KB	4 KB	16 MHz
Due	AT91SAM3X8E	54	12	12	512 KB	96 KB	-	84 MHz

Arduino Shields

Wouldn't it be cool if your Makerspace had its own Arduino compatible shield? Of course it would! Maybe your Makerspace is really into FIRST robotics, or screen-printing, or whatever, but there isn't a shield that meets your specific needs. By developing your Makerspace's own expansion shield, it provides a common project for the Makers as well as setting your space apart from the others.

As Arduino has predetermined I/O and spacing (Figure 5-50), it is easy to design a shield without having to do a lot of design work. If you reference the outline of the original (nonMega), Arduino from the previous section and remove all of the circuitry, what is left is the bare outline of the board. This outline should serve as a starting point for your design. Although it is not necessary for the shield to conform to the Arduino's board outline and mounting hole placement, it serves as a reference for establishing the physical boundaries of the Arduino and how it might affect the design of the shield.

I/O Considerations

There are a series of inputs and outputs that you should take into consideration during shield design. The first of which is the reset button. On most stock Arduinos, the reset button is located near the center of the board. When a shield is stacked on the Arduino, the reset button is covered and the function is lost. Shields can combat this problem by including a reset button in their design and positioning it near the edge of the board for easy access. The next notable feature is the ISP header. This header is located in the same position on all Arduinos and makes it possible to upload the Arduino boot loader to the microcontroller or simply programming over ISP. Because this header doesn't move across designs, it is an easy feature to include on a shield. Simply add the 6-pin header to the shield and use a female header positioned on the bottom of the board to make the connection. Finally, the last feature to note is the use of the SCK pin, or pin 13. This pin is widely used as a status indicator because all Arduinos have an LED attached to the pin. Ensure that your shield takes the use of the pin into account and does not tie it to a piece of hardware that should be switched on and off during programming, like a motor controller.

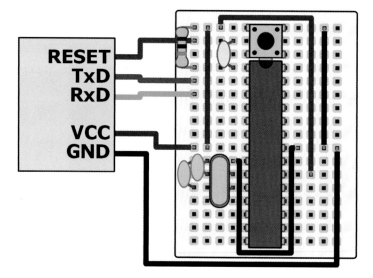

Figure 5-49. *Wiring diagram for the breadboard Arduino.*

Power Considerations

Arduino gets its power from an external source and is converted to either 3.3 VDC or 5 VDC via an onboard regulator. If the source is a USB port, the current is limited to 500 mA and can be substantially higher when using a dedicated power supply. Table 5-23 illustrates the voltage and current capability of each regulator.

If you were to include a motor driver, relay, or other power-hungry device onto Arduino's regulated power bus, you might run into some problems. Most of these regulators regulate their voltage output by burning excess voltage off as heat; the more current that flows through the regulator, the hotter it gets. If too much current is drawn, the regulator will begin to drop the output voltage or turn it off completely as a safety measure.

During your design phase, create a power budget or table of components that consume power. If the shield will be powering an excessive amount of devices, it might be necessary to give it its own regulator. A power budget illustrates the power consumption of all of the electronic components attached to the power bus. When added up, these power consumptions should be less than the maximum power output from the regulator by a small margin. This margin accounts for any potential power spikes and should help to prevent brownout situations.

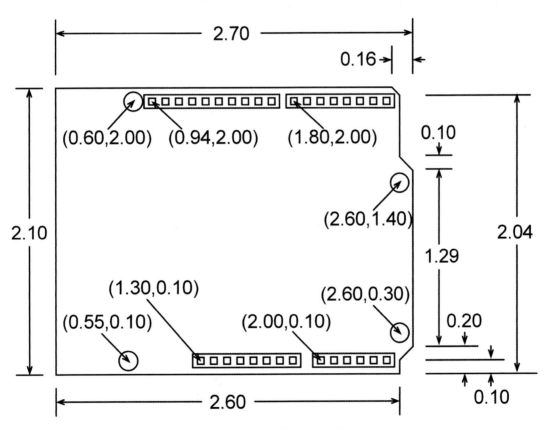

Figure 5-50. *Arduino dimension reference.*

Table 5-23. Arduino power output capability

Board(s)	Regulator	Voltage	Current
Uno, Leonardo, Mega	NCP1117	5.0 V	1.0 A
Uno, Leonardo, Mega	LP2985	3.3 V	0.15 A
Due	LM2734	5.0 V	1 A
Due	NCP1117	3.3 V	1 A
Mini	LP2985	5.0 V	0.15 A

Space Constraints

It is easy to disregard 3D constraints when you spend all of your time working on a 2D circuit board design (Figure 5-51). This is especially important with shield designs because the exposed through-hole component pins can contact some of the Arduino's components. The primary areas of concern are around the USB jack, power connector, and the ISP header. Using Arduinos that utilize a USB Mini connector eliminates the USB interference problem but requires the user to employ only Arduinos with that type connector, and thus doesn't allow for universal compatibility, so design for the worst case. The power connector is pretty unavoidable. Some Arduino variants use a JST battery connector in lieu of the bulky 2.0 mm barrel jack, but as with the USB problem, it is best to design with the standard connector in mind. Lastly the ISP header can actually be a nice design feature to embrace, allowing for another interface to power, SPI, and reset.

Figure 5-51. *Consider the space requirements of all of the components between the headers while you design your shield.*

Project: Arduino RFID Power Control

One of the problems you might encounter when running a Makerspace is keeping tabs on who has access to equipment. Tools like the laser engraver, 3D printer, and power tools require the operator to understand how to properly operate the machine and how to know when something is going awry. This project is designed to provide a starting point for an RFID-based lockout system (Figure 5-52) that provides power access to equipment. The example sketch sets access to only one ID but can easily be reworked to accommodate multiple IDs and timed access.

Figure 5-52. *The lockout circuit.*

Arduino sits at the heart of this project, providing the link between the card swipe and the power switch. A more permanent solution would be to utilize a less expensive Arduino variant that is more suited for embedded projects. Boards like the SMDuino are designed specifically for embedded projects by utilizing a small form factor and minimal components while maintaining Arduino-like functionality.

This project requires that safety glasses be worn throughout its entirety.

Materials

You can find the files for this project at Thingiverse.com (*http://bit.ly/19K4fYb*).

Materials List		
Material	**Qty**	**Source**
Electronics enclosure (optional)	1	Electronics supply store
Female-to-female jumper wires	9	Electronics supply store
0.1 in male header	9	Electronics supply store
Small gauge stranded wire	6 ft	Electronics supply store
PowerSwitch Tail II	1	*http://www.powerswitch tail.com/*
Adafruit NFC shield	1	*http://www.adafruit.com/*
20 × 4 character LCD	1	Electronics supply store
4.7 kOhm resistor	1	Electronics supply store
4-40 × 1 in machine screw	8	Hardware store
4-40 flat washer	8	Hardware store
4-40 nut	8	Hardware store
4-40 × 1/4 in spacer	8	Electronics supply store

Makerspace Tools and Equipment
Laser engraver
Soldering iron
Solder
Safety glasses

Procedure

Step 1

Begin by programming the Arduino with the sketch shown in Example 5-1. This program is meant to act as a starting point for securing your machinery.

Step 2

Using a soldering iron, solder a series of male headers into the second row of pads located on the Adafruit NFC Shield and into the LCD. Use these to connect the Arduino to the LCD by using male-to-male jumper wires. Following the wiring diagram (Figure 5-53), use the male-to-male jumper wires to connect the top of the NFC shield to the LCD. Solder the 4.7 kOhm resistor between pins 1 and 3 on the LCD. Connect a pair of control lines to pin 10 and GND on the Arduino. These will be used to control the PowerSwitch Tail.

Step 3

Cut out and assemble the acrylic enclosure (Figure 5-54). Alternatively, you can use and modify a conventional electronics enclosure to support the components. Mount the LCD using the 4-40 fasteners and spacers. Mount the Arduino to the base plate using another set of fasteners and spacers. Place the NFC shield onto the Arduino and plug in the power. Follow the directions for programming the "master" card. Once the device is programmed and functional, connect the control lines to the PowerSwitch Tail and the desired piece of equipment and you are good to go!

Example 5-1. Basic Arduino RFID lockout

```
// Based on the Adafruit example code from their NFC Shield Library

#include <Wire.h>
#include <Adafruit_NFCShield_I2C.h> // ❶
#include <LiquidCrystal.h> // ❷
#include <EEPROM.h> // ❸

const int IRQ = 2; // ❹
const int RESET = 3;
const int powerPin = 10;  // ❺

Adafruit_NFCShield_I2C nfc(IRQ,RESET);

LiquidCrystal lcd(12,11,6,5,4,3);

boolean nfcReady = false;
uint8_t uid[7]; // ❻
uint8_t uidLength;
```

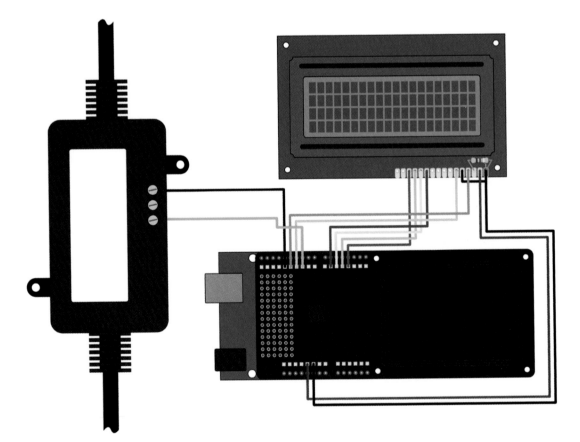

Figure 5-53. *Wiring diagram for Arduino RFID power control.*

```
boolean accessEnabled = false;
boolean accessAllowed = false;
boolean powerEnabled = false;

void setup() {
  nfc.begin(); // ❼
  lcd.begin(16,4); // ❽

  pinMode(powerPin, OUTPUT); // ❾
  digitalWrite(powerPin, LOW);

  int nfcVersion = nfc.getFirmwareVersion();

  if (!nfcVersion) { // ❿
    lcd.setCursor(0,0);
    lcd.print("NFC comm. error!");
    lcd.setCursor(0,1);
    lcd.print("Halting...");
    while(1);
  }
  else { //display startup text
    lcd.setCursor(0,0);
    lcd.print("RFID Power Lock");
    lcd.setCursor(0,1);
    lcd.print("NFC Fw: ");
    lcd.print((nfcVersion>>16) & 0xFF, DEC);
    lcd.print(".");
    lcd.print((nfcVersion>>8) & 0xFF, DEC);
    lcd.setCursor(0,3);
    lcd.print("Starting up...");
  }

  nfc.setPassiveActivationRetries(0xFF);
  nfc.SAMConfig();
  delay(2000);
  lcd.clear();
  checkAccess();
}

void loop() {
  nfcReady = nfc.readPassiveTargetID(PN532_MIFARE_ISO14443A, &uid[0], &uidLength); // <⓫>

  if (nfcReady) {
    if (accessEnabled == false) { // ⓬
      lcd.clear();
      lcd.setCursor(0,0);
      lcd.print("Read card!");
      lcd.setCursor(0,1);
      lcd.print("UId: ");
      for (int i = 0; i < uidLength; i++) {
        lcd.print(uid[i], HEX);
        EEPROM.write(i,uid[i]);
      }
      lcd.setCursor(0,2);
      lcd.print("Swipe card again");
      lcd.setCursor(0,3);
      lcd.print("to enable Power.");
```

```
      accessEnabled = true;
      // Wait 1 second before continuing
      delay(1000);
    }
    else { // ⑬
      accessAllowed = true;
      lcd.clear();
      lcd.setCursor(0,0);
      lcd.print("Swipe card to");
      lcd.setCursor(0,1);
      lcd.print("toggle power...");
      for (int i = 0; i < uidLength; i++) {
        if (byte(uid[i]) != EEPROM.read(i)) {
          accessAllowed = false;
        }
      }
      if (accessAllowed == true) {
        if (powerEnabled == false) {
          digitalWrite(powerPin, HIGH);
          lcd.setCursor(0,3);
          lcd.print("Power is ON.");
          powerEnabled = true;
        }
        else {
          digitalWrite(powerPin, LOW);
          lcd.setCursor(0,3);
          lcd.print("Power is OFF.");
          powerEnabled = false;
        }
      }
      else {
        digitalWrite(powerPin, LOW);
        lcd.setCursor(0,3);
        lcd.print("ID not allowed.");
      }
      // Wait 1 second before continuing
      delay(1000);
    }
  }
  else
  {
    // ⑭
    lcd.clear();
    lcd.setCursor(0,0);
    lcd.print("System timed out");
  }
}

void checkAccess() { // ⑮
  byte inByte = EEPROM.read(0);
  if (inByte == 0) {
    lcd.clear();
    lcd.setCursor(0,0);
    lcd.print("Swipe card to");
    lcd.setCursor(0,1);
    lcd.print("set access key...");
    accessEnabled = false;
```

```
    }
    else {
      accessEnabled = true;
      lcd.setCursor(0,0);
      lcd.print("Swipe card to");
      lcd.setCursor(0,1);
      lcd.print("toggle power...");
      lcd.setCursor(0,3);
      lcd.print("Power is OFF.");
    }
}
```

❶ Super handy NFC library from Adafruit. Download it from *https://github.com/adafruit/Adafruit_NFCShield_I2C*.

❷ This brings in the LCD library, which is included with Arduino.

❸ Use this library to write data to Arduino's nonvolatile memory.

❹ Specifies which Arduino pin (2, in this case) is used for an interrupt pin to communicate with the NFC reader.

❺ This determines which pin controls the power supply.

❻ This buffer stores the UID value that's read off the NFC tag.

❼ Initializes the NFC shield.

❽ Initializes the LCD display.

❾ Configure the power control pin as an output and then turn it off.

❿ If the NFC shield can't be initialized, display an error.

⓫ This flag indicates whether a card is present.

⓬ Checks to see whether a "master" card has been established.

⓭ If this is the master card, toggle the power.

⓮ This code gets run if the NFC shield timed out while waiting for a card to appear.

⓯ This sets the "master" card ID for locking and unlocking. You'll need to reset it with an EEPROM wipe (*http://bit.ly/13U8eJl*).

Figure 5-54. *This project should serve as a starting point for securing your equipment and is meant to keep the honest people honest.*

2D Design and the Laser Engraver

The laser cutter/engraver might be the single most useful piece of equipment in the Makerspace (for the sake of ease, I'll refer to it simply as the "laser cutter" from this point on). Its ability to quickly create usable objects from many modern materials makes it the go-to machine for most projects. These machines interface directly with a computer, and you can output files to it as if it were a conventional printer. You use software to produce *raster* and *vector* designs that are translated into a series of paths for the laser. Raster refers to images made up of pixels that result in the machine engraving the image and its contrast onto the material. Vector designs contain line paths that result in the machine replicating the path with the laser, thus cutting the material.

Because laser cutters can handle a wide range of material sizes and types, it is important to pro-vide adequate storage. Most large machines have wheeled stands with storage compartments, yet their capacity is limited. I recommend that you use a half-sized filing cabinet with two or more drawers. This makes the separation of new and scrap materials convenient and helps reduce waste materials piling up around the machine.

Like the 3D printer, laser cutters require a computer, power, ventilation, and storage. The host computer communicates with the machine by using a print driver and therefore does not need to be located in the same proximity. Attached to the laser cutter is a significant ventilation system. This system is designed to remove the often harmful vapors produced while the machine is in operation and can take up a considerable amount of floor space. Because of this ventilation constraint, direct access to the exterior of your facility must be made available, unless you are using a ventilation system designed for indoor use.

Some laser engravers have the ability to connect to the network so that multiple computers can access it at once. While this is a convenient way to give access to the machine, it greatly increases the potential for problems. Restricting print access to the engraver through one dedicated computer helps to ensure its proper operation and will ultimately make managing it easier.

Table 6-1. Freely available 2D CAD software

Software Title	Developer	Platform	Link
Draftsight	Dassault Systèmes	Linux, OS X, Windows	http://www.3ds.com/products/draftsight
LibreCAD	Community	Linux, OS X, Windows	http://librecad.org
QCAD CE	RibbonSoft	Linux, OS X, Windows	http://www.ribbonsoft.com

Facilitating Design by Using 2D CAD

When entering the world of two-dimensional computer-aided design, or CAD, you will encounter an endless supply of software options. Some Makers prefer to use software that focuses primarily on graphic design, such as Illustrator or CorelDraw from Corel Corporation. Although these applications work perfectly fine with a laser cutter, they lack the mechanical drawing tools found with conventional CAD. CAD is specifically tailored to produce working drawings by means of tools that really make prototyping easy and exporting to the laser cutter a snap.

In the past, physical objects were created from hand-drawn blueprints that detailed every necessary design feature. This typically involved a three-view drawing that illustrated the front, right-side, and top of an object, and used drawing conventions to illustrate features such as dimensions, internal holes, threading, and so on. Now that software has positioned itself between the pen and the hammer, designs can be created quicker and with higher accuracy than ever before, and happily there are quite a few freely available CAD options that your Makerspace can adopt. Table 6-1 lists three of them.

The 2D Design Environment

Regardless of the choice of software, there are a series of standard design tools with which you can create your virtual design. With these tools, a designer can transfer pen and paper ideas to a medium that is directly compatible with the laser engraving and cutting process. The laser cutter's ability to quickly convert these designs into physical objects enables a working environment with near limitless bounds for your imagination.

The first setting that should be set within the CAD software is the default unit of measure. Even though CAD is capable of operating in a unit-less environment, it is best to establish this default unit set and make it standard across the Makerspace. Doing this ensures less wasted material and time when trying to convert between bases.

Your Makerspace should also implement the concept of a read-only template that is used specifically for sending engraving and cutting jobs to the laser engraver. This *default template* helps to reduce the chance of improperly configuring the engraver and wasting material in a failed print job. The final tip for configuring the CAD software is to make the default path for opening files a network accessible drive. This drive can then act as a "drop box" to which your Makers can post files for engraving and cutting and will help reduce the amount of time required at the laser engraver's computer. This machine will see a lot of traffic in your Makerspace, so every attempt should be made to expedite the process.

Drawing Tools

The CAD software's drawing tools make it possible for you to create almost any design. You can work with elements as simple as a line and as complex as a multisegmented Bezier curve to define the shape and size of your project. With these tools, you can create designs that the laser cutter will reproduce as burns and cuts to a slew of materials, and understanding their function will make the design process that much quicker. Most CAD software suites implement a common set of tools meant for drawing, object creation, and workspace manipulation. Like any standard program, you can select these tools with the simple click of a mouse or through the faster method of using keyboard commands and hot-keys.

Arc

Arcs are can be defined in two ways: by the center of the arc and its endpoints, or by defining the start point, middle, and endpoint. The resulting segment can be altered by selecting any one of the three nodes and altering its position.

Circle

Circles are defined in two ways: by the position of their center, or as a tangential shape between two points. After the circle is created, you can alter it by changing the location and size of the radius.

Ellipse

You create ellipses in the same fashion as circles, although there are two radius features that you can alter.

Line

The line tool creates a two-point line within the drawing space. The line itself is defined by two *x/y* coordinates that determine the start and end of that line segment. You can change these coordinates by either selecting one of the endpoints and altering the

coordinate values or by moving the entire line segment.

You can also use lines to construct whole polygons. By creating multiple lines and aligning or connecting their endpoints, you can convert the resulting shape into a polygon. In the vector drawing software, this command is commonly referred to as a *JOIN*, and in CAD you can complete it by using the PEDIT command.

Point

Points define an individual *x/y* coordinate. These are quite useful when defining the shape of a spline using predetermined coordinate points. A good example of this is creating an airfoil shape based on a series of *x/y* intersects. You place the points in the workspace, and then you can alter them by selecting the point and changing its location.

Polygon

The polygon tool is useful for creating complete polygon shapes. When the tool is selected, you can input the desired parameters to define the number of sides, whether it is defined as inscribed or circumscribed to a circle, and the angle of each segment. After the shape is created, you can alter it by individually selecting each node or by separating the polygon into individual line segments using the *SPLIT* or *EXPLODE* command.

Polyline

You use the polyline tool to create Bezier paths defined by a string of line and arc segments. You position these segments with each subsequent click of the mouse, and you can switch between line and arc typically by right-clicking your mouse. This tool is useful when attempting to draw an arc along a continuous path.

Rectangle

Although you can create rectangles by using the line tool and joining the individual line segments, the rectangle tool automatically creates a shape whose line segments are already attached. This constraint on the endpoints makes it possible for you to initially create the rectangle with specific dimensions and alter them after its creation. Thus, editing the length of one side changes the length of the opposite side.

The line segments in the rectangle can be separated into their individual segments by using the SPLIT command in the vector drawing software and the EXPLODE command in CAD.

Spline

Splines are mathematically based Bezier curves that are defined by a series of handles. After creating a series of connected splines, you can alter the shape of the curve by selecting each node, or endpoint, and changing the position of the handle. The more you pull the handle away from the node, the more it distorts the curve. This tool is especially useful when outlining an image or recreating an existing organic design, such as a tree. Because the spline tool is mathematically based, you can scale its attributes and manipulate them later.

Workspace and Object Manipulation Tools

The workspace is the area in which you draw. In this limitless space, you can create anything from items of submillimeter size to objects as big as a house. Because of this vastness, it is important to consider the limitations of your tools and materials when designing your project.

Array

You use the array tool to create multiple copies of a specific element. You can configure it to create a linear or polar array. The linear array is used to produce a grid of copies. You configure it by setting the quantity of rows and columns as well as their offsets. You use the polar array function by defining the center point of the polar array, the quantity of copies, and their angle separation.

Dimensioning

Using the dimensioning tool, you can analyze the actual dimensions between two points and verify it with a visual readout. This series of tools can be used to determine linear or angular distance, angle, and radius, and you can also use it to edit and verify the correctness of your design.

Grid

The grid tool creates a virtual reference grid that extends from the origin at the unit and delineation of choice. It is commonly used as a visual guide for the placement and scale of drawing elements.

Grid Snap

After the grid unit and delineation has been specified, you can use the grid snap feature to snap element features to the grid or act as a reference for the creation of new elements. This tool helps to maintain the organization and scale of your drawing.

Layers

Drawing in CAD is inherently similar to drawing with a pen and paper. You can illustrate your design on two dimensions, but if you need to draw an overlapping element, you would need another sheet of paper. CAD accommodates this by using layers. Each layer acts as a new drawing that overlaps each successive layer and helps to alleviate confusion when viewing complex designs. You can use layers to distinguish drawing features such as construction lines from the rest of the drawing. Using the layer manager, you can create as many layers as you need, change their visibility, color, alterability, and even change whether they are sent to the printer.

Move

You can use the move tool to position entire elements or specific parts of an element. To move an object, select it and activate the move tool. You can then translate it by clicking and dragging or by entering the desired coordinate location.

Object Snap

You can define objects by several attributes, points being one of them. You can then use these points to snap other elements to these points. You can configure the snap tool to snap elements to the endpoints, midpoint, tangent, center, overlapping intersections, and perpendicular intersections. This is a useful tool when nesting drawings in preparation for cutting or when fit-checking joints.

Object Tracking

Object tracking utilizes the defining features of an element to act as a reference point for creating new elements or placing existing ones. This tool creates a reference line that extends from the selected feature, which you extend to any desired length. In CAD, the length of the line can be set while the tool is active.

Orthographic/Polar Tracking

Similar to object tracking, you can configure polar tracking to create a reference line from any point that snaps to the desired polar angle. This snapping action simplifies the creation of perpendicular or properly angled lines without having to calculate distances. This tool is commonly configured to snap to 30 or 40 degrees.

Scale

You use the scale tool to change the size of one or more elements based upon an origin point and a scale factor. When using this tool, highlight all of the elements you want to scale, select the scale tool, and enter the origin and factor. As the use of the metric system becomes more and more popular, so does the demand to scale objects between systems. The most common occurrence is scaling between the Imperial and metric systems. Table 6-2 illustrates some of the more common units and their scale multipliers.

Table 6-2. Scale multipliers

Unit	Imperial (in)	Metric (m)
1 ft	12	0.3048
1 in	1	0.0254
1 mil	1×10^{-3}	2.54×10^{-5}
1 m	39.37	1
1 mm	0.03937	0.001
1 micron	3.937×10^{-5}	1×10^{-6}

This table comes in handy when importing a drawing that was scaled in millimeters and your engraver is configured for inches. Using the scale tool, highlight the drawing and set the scale factor to 0.03937.

Commands and Hot-Keys

It might appear that using the mouse to find a tool is the quickest option for selecting it. As your proficiency for the use of the CAD software improves, you will find that keyboard commands and hot-keys are considerably quicker. Table 6-3 lists some of the more common commands and hot-keys that will help expedite the creation of your drawing.

Table 6-3. CAD commands and hot-keys

Tool	Command	Hot-key	Function
Arc	ARC		Selects the arc tool
Circle	CIRCLE		Selects the circle tool
Explode	EXPLODE		Separates a polyline into separate elements
Help	HELP	F1	Opens help window
Line	LINE		Selects the line tool
Object snap	OSNAP	F3	Toggles object snap
Offset	OFFSET		Sets the offset of a new element from an existing one
PEdit	PEDIT		Combines individual elements into a polyline
Rotate	ROTATE		Selects the rotate tool
Spline	SPLINE		Selects the spline tool
Trim	TRIM		Defines the cursor as a cutting tool for element removal
Unit	UNITS		Opens the unit system dialog

Joints

Although the laser cutter is a stellar machine for producing two-dimensional objects, there are design elements that you can use to construct three-dimensional objects. In methods very similar to wood working, you can create the desired object by puzzle-piecing together a series of sides and securing them in place with adhesive.

These connect two or more pieces together to form an assembly. The objective of the joint is to provide enough strength at the interface to handle the expected stresses. Because these stresses can vary greatly, there are quite a few options for producing an acceptable joint. Using an optimized joint in your assembly will help to improve its strength and will make for a much better final product. The joints in this section work well with the limitations of the two-dimensional cutting capability of the laser cutter and can be implemented into virtually any design.

Butt

> The easiest method for producing a joint is to butt two pieces of material together with a small amount of adhesive (Figure 6-1). This type of joint is not very strong because it doesn't support the material from torque and lateral loads, but it works well with small and light designs that won't experience large forces.

> Variations to the butt joint include the *rabbet* and *lap* joints (Figure 6-2). These joints are designed to increase the surface area at the mating interface, which greatly improves the joint's strength.

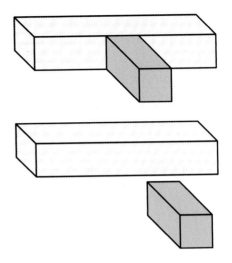

Figure 6-1. *Butt joint.*

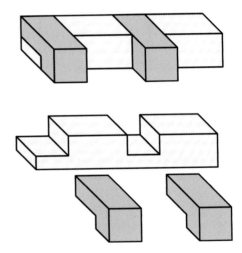

Figure 6-2. *Rabbet and lap joints.*

Finger

> The finger joint is the best solution for most designs because of simplicity and its ability to produce a very strong coupling without wasting material (Figure 6-3). Because the mating interfaces are mirrored images of each other, they can be used to produce a 0-degree or 90-degree mate and can even be cut simultaneously if the two elements are overlapped in CAD.

> Beause they don't contain any complex design elements, finger joints are quite easy to

create. You can create them by using two different methods. The first method utilizes a predetermined size for the finger that is no less than the thickness of the material. Starting at the center of the mating side, copy and paste alternating rectangle squares until one square overlaps the end of the line. Trim off the excess and repeat the procedure for the other side. The second method requires dimensioning the length of the side and dividing it into segments. The length of the segments then becomes the size of the finger, which you can create on the side without any residual overlap.

A variation of the finger joint is the tapered finger joint (Figure 6-4) which is used exclusively to mate two pieces of material together to increase the overall length.

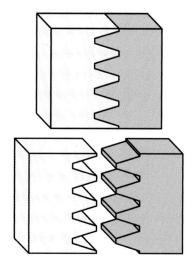

Figure 6-4. *Tapered finger joint.*

Ensure that you remove any protective covering when measuring material thicknesses. A small amount of protective coating can result in an oversized hole.

Mortise and Tennon

Similar to the finger joint, the mortise and tennon joint utilizes a pocket mortise that provides greater lateral support (Figure 6-5). A side benefit of this joint is that it is sturdy enough to act as a temporary interface without the use of adhesive, allowing for the parts to be mated and separated repeatedly for dry fitting.

All that is needed to draw a mortise and tennon joint is a standard dimension. This dimension is used to dictate the length of the mortise and tennon, with width being the thickness of the material. Ensure that there is enough support material around any side of the mortise to support the tennon, and a good rule of thumb is to make it no less than the thickness of the material.

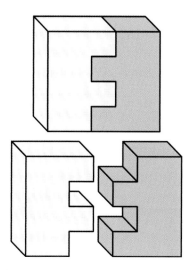

Figure 6-3. *Finger joint.*

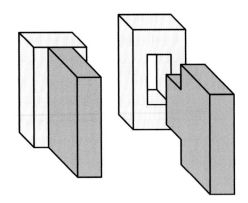

Figure 6-5. *Mortise and tennon joint.*

Measure your material before you cut. Manufacturers use many different sizes to represent a "standardized size." For example, 1/4 in acrylic can measure 0.222 in as well as 0.205 in. This small difference can result in a loose-fitting and weak joint.

Nested Fastener

The nested-fastener joint has become quite popular in Maker designs because it can hold material together without adhesive (Figure 6-6). Expanding the mortise and tennon concept, the nested-fastener joint supports a nut perpendicular to its interface on the tennon side and aligns it with a fastener on the mortise side. To assemble the joint all that is needed is to mate the two pieces, place the nut in the cutout, and then secure the pieces together with an appropriately sized bolt.

To produce a nested-fastener joint, begin by dimensioning your components. The nut should be nested so that two of its faces sit perpendicular to the host material. Use a pair of calipers to measure the width of the nut and add a 10 percent margin to your measurement. This will make it easy for you to place the nut into the slot. The second dimension is the length of the fastener. Just as with the nut measurement, measure the length and width of the fastener and add 10 percent.

When drawing the modified mortise, take its length, let's assume 1 in, and divide it into thirds. The first and last box will be the mortise and the center box will support the fastener. Next, bisect the center box and draw a circle with the diameter of the threaded portion of your fastener.

To draw the modified tennon, use the same length as the mortise, in this case, 1 in. Beginning at the center of the mortise, draw a reference line the same length as the length of the fastener. Offset two lines at half of the diameter of the fastener and complete the rectangle. Finally, position the nut so that it is approximately 1/8 in. from the end of the rectangle and trim the construction lines.

If completed correctly, the joint should mate together and allow for the fastener to be passed through the center of the tennon, which you can then secure in place with the embedded nut.

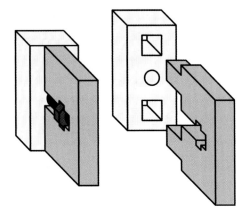

Figure 6-6. *Nested-fastener joint.*

The nut has the tendency to fall out of the cutout, especially when working in tight positions. A simple piece of tape placed over the cutout after the nut has been nested is a simple and effective solution.

Project: Creating a Default Template

Although this project is not very complex, its usefulness is unparalleled. The laser cutter is considered just another printer to which you can output, and your drawings can be difficult to accurately position within the printer's bed area. By having a preconfigured default template (Figure 6-7) that replicates the exact bed area, you can easily position jobs and send them to the cutter.

Procedure

You can find the files for this project at Thingiverse.com (*http://bit.ly/19K4DFY*).

Step 1

Begin by determining where the *x0/y0* position is for your engraver. Some use the upper left and some the upper right. This is important because it dictates the position in which you construct the template. Open a blank file in your CAD software and create a new layer titled "bed." Use the rectangle tool to draw the perimeter of the bed area with respect to the *x0/y0* coordinate.

Step 2

Each laser engraver manufacturer uses a print driver that is specific to the system (Figure 6-8). This driver is used in lieu of the traditional print dialog and makes it possible for you to send settings to the printer in addition to the drawing.

To configure the driver, click File and then Print. Set the driver for the appropriate bed size, no margins, and 1:1 scale. Power up the engraver and send the drawing to the engraver. The specific power/speed settings do not matter at this point because this is just a test. If your laser cutter is equipped with a

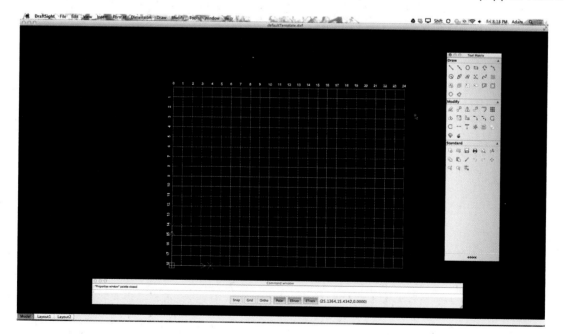

Figure 6-7. *The default template.*

laser pointer feature, enable it; otherwise, this will take a bit of guess and check. Raise the lid and press Go. The laser cutter should begin tracing the outline of the rectangle. Visually note if there is any deviation from the extent of the bed area and approximate the distance. This distance is the "offset" that you need to enter into the print driver to correct the problem. Some applications don't like printing without margins, so it tends to mess things up.

Step 3

Enter your approximate offset into the print driver configuration window and repeat the previous steps until the engraver traces the exact perimeter around the bed.

Step 4

Draw a 1 in (or whatever unit you wish) grid pattern across the build area; this will assist with part placement within the template (Figure 6-9). Above each row and to the left

of each column, label their measurement by using the text tool.

Step 5

Freeze the layer to prevent unintended alterations and disable the print function for it. This will prevent the template from being unintentionally sent to the laser cutter when printing. Highlight the original "0" layer and save the file. Close the file within your CAD software and locate it within your operating system. Change the file properties to "read-only." This prevents anyone from accidentally messing with the settings and altering the template.

This template should now be the only file your Makerspace uses when printing to the laser cutter. Instruct your Makers to open their CAD files and copy and paste them into the template to ensure proper printing and the best possible results.

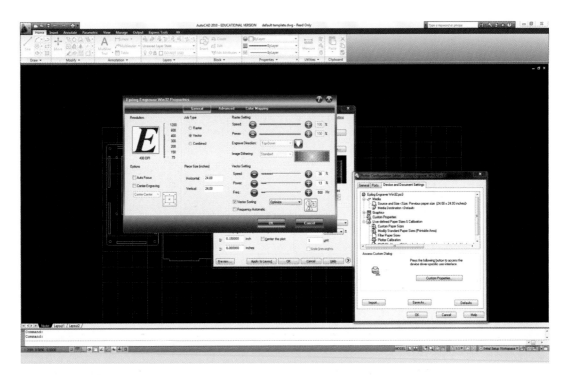

Figure 6-8. *Epilog driver properties.*

Laser Fundamentals

The laser cutter is a machine that manipulates and focuses a high-power laser beam onto a piece of material following a two-dimensional path generated by a computer. This makes it possible for you to either cut or engrave certain materials with very little waste. The result is a blade-less engraving and cutting tool that has virtually limitless application.

About the Laser

At the heart of the machine is a beautiful laser. This apparatus is designed to emit a high-power beam of amplified light through a series of mirrors and lenses so that it can be used to subtract material in a controlled manner. These tubes are so high powered that they need to be actively cooled. Depending on the manufacturer and design, some tubes use a liquid cooling system, while others use air.

With the carbon dioxide (CO_2) laser, which is the most common for commercial laser engravers, laser light is generated inside of a gas discharge tube filled with a mixture of carbon dioxide, helium, hydrogen, and nitrogen. One end of the tube contains a fully reflective mirror and the other a mirror that is partially reflective. Energizing the gas inside the tube excites the atoms. When a greater percentage of the population has reached this excited state, a population inversion occurs, which improves the overall efficiency of the system. As the atoms begin to fall back to their ground state, they release an intense photon at a wavelength specific to the electron's energy at the point of emission, which is around 10.6 microns for CO_2. This light energy is then reflected within the tube, gaining energy as it bounces back and forth between the mirrors, causing more atoms to release photons. When the light has amassed enough energy, it escapes through the partially reflective mirror in the form of a coherent laser beam.

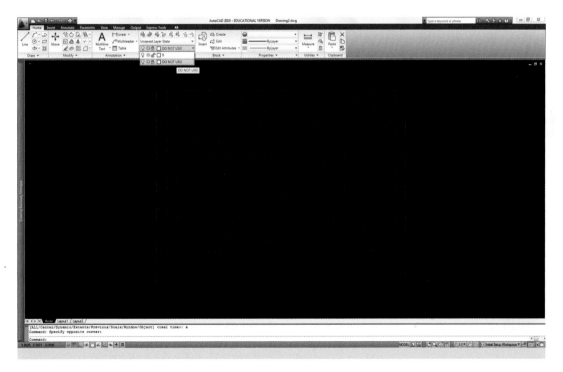

Figure 6-9. *Set the grid pattern to a locked/nonprinting layer. This way, you can draw on top of the pattern and it won't be sent to the laser.*

Choosing the Right Laser

As with any tool, it is important to choose the right laser cutter for the job, and there is no exception with a machine this expensive. Laser cutters vary greatly in price depending on manufacturer, laser power, lens quality, and bed size, and all of these factors need to be considered before taking the plunge. Because every Makerspace has different expectations and requirements, having a good understanding of the capabilities of each laser type will help make the decision easier.

Generally, the more powerful the laser, the more materials and thicknesses it can handle. Manufacturers generally scale the cost of their machines both on the wattage of their laser as well as the size of the bed. The decision regarding which is more easily sacrificed depends on the focus of your Makerspace. More wattage means that you can work with thicker, denser materials, whereas a larger bed accommodates bigger designs. Table 6-4 compares three of the different laser types and the materials that they can cut and engrave.

Table 6-4. Available laser engraver wavelengths

Laser Type	Wavelength (micron)	Power (watt)	Application
CO_2	10.6	30–120	Engraving and cutting many organic and inorganic materials as well as some metals
CO_2	9.3	30–50	Specialized for engraving and cutting specialized materials such as PET, photopolymer, and polyimide
Fiber	1.06	10–50	Marking metals and some plastics without surface coatings

In general, 30- to 40-watt laser cutters can cut materials up to 1/4 in. in thickness. Thicker materials can be cut by performing multiple passes, but depends on their density. Machines in the 50- to 60-watt range can cut material up to 3/8 in thick and can cut thicker using the multiple-pass method. Table 6-5 presents common materials

with which you can use a laser cutter, and whether they can be engraved, cut, or both.

Material Types

The types and thicknesses of materials that can be engraved and cut using the laser engraver depend directly on the machines laser type, power, and lens assembly. These materials are also chosen due to their lack of harmful compounds. These compounds might be inert when the material is in its normal state, but can be released during engraving and cutting.

Make sure you know the chemical content of the materials with which you operate the machine. Materials like PVC and others that contain hydrogen chloride and vinyl chloride release chlorine gas when vaporized. This is incredibly dangerous and should never be attempted!

Table 6-5 shows a table of common laser-able materials and the required laser type.

Table 6-5. Common materials that a 10.6-micron machine can engrave and/or cut

Material	Engrave	Cut
Plastics		
ABS	YES	YES
Acrylic	YES	YES
Nylon	YES	YES
Wood		
Hardwood	YES	YES
Softwood	YES	YES
Plywood	YES	YES
Fabrics		
Cotton	YES	YES
Satin	YES	YES
Silk	YES	YES
Vinyl	YES	YES
Polyester	YES	YES
Felt	YES	YES

Material	Engrave	Cut
Leather	YES	YES
Paper		
Paper	YES	YES
Vellum	YES	YES
Mat board	YES	YES
Metals [a]		
Anodized aluminum	YES	NO
Iron	YES	NO
Steel	YES	NO
Painted brass	YES	NO
Natural Materials		
Glass	YES	YES
Cork	YES	YES
Stone	YES	NO
Latex rubber	YES	YES

[a] Metals require specialized coatings to enable engraving

Lens Types and Focusing

There are multiple lenses and mirrors that work together to divert the high-power beam produced from the laser to the material. When the beam enters the lens assembly on the carrier, it is first reflected at a 90-degree angle and focused onto the material. The engraving and cutting capability of the engraver depends both on the power of the laser and the ability to focus the beam on the workpiece.

Lens Types

There are a variety of lenses you can install into the machine's lens carrier. These lenses are responsible for taking reflected laser light from the tube and focusing it onto the desired material. The lens itself is contained within an aluminum structure (Figure 6-10) that aligns the lens with the incoming laser light. Most manufacturers provide easy access retainers that make it simple to swap lens types and expose the optics for cleaning.

Figure 6-10. *Lens assembly.*

Standard

The standard lens is designed to produce good all-around results. Most of the engraving and cutting specifications set by the manufacturer pertain to this lens.

Deep Cut

These lenses feature a long, focused beam that is optimal for engraving recessed areas as well as cutting thick material. Because the beam is not highly focused, the resulting engraving resolution is relatively low and produces a wide kerf when cutting.

High Resolution

"High-resolution" or "high-definition" lenses are designed to produce a very small spot size, ~0.003 in. These lenses work best when used for high-resolution engraving and cutting thin materials.

Focusing

You focus the laser by using one of two main methods: electronic or mechanical. Both require an initial determination of your laser's focal point. Every machine should be equipped with either a focus jig or autofocus feature. To focus your laser, simply position the desired material onto the cutting bed and free the *x/y* axis. You do this by pressing the Focus button on the machine's control panel, or through other means, depending on the manufacturer. Slowly adjust the height of the z-axis until the focus jig just touches the material's surface. Remove the jig and reset the axis. It is important to focus the machine every time it is used. This is one of the more forgotten steps when operating the machine.

If you do not know the focal length of your engraver's lens or the jig needs to be adjusted, you can achieve this distance through simple trial and error. Send a single line drawing to the machine and systematically conduct a series of tests to determine the appropriate distance by analyzing the width and quality of the cut (Figure 6-11).

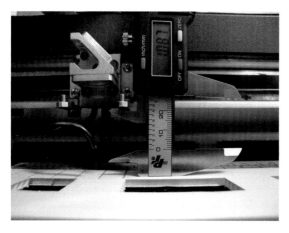

Figure 6-11. *Determining focus.*

The Print Driver

The print driver bridges the gap between the laser cutter and your software design. It dictates the method by which your drawing will control the speed, power, and motion of the laser, as well as a slew of operational parameters. In addition, you can configure the driver to utilize predetermined material engraving and cutting settings as well as accepting custom configurations.

General Settings
Build Area

The build area boxes contain the length and width measurements of your engraver. These measurements are then used to produce a custom sheet size in your CAD software's print dialog and is used to define the limits of your workspace. Stock print drivers

come preloaded with a series of presets that are often designed for more than one type of machine and can contain bed dimensions different from your machine. Simply adjust the dimensions to fit your machine and overwrite the preset values. This ensures that your machine utilizes the full extent of the bed area and reduces the chances of unintended problems.

Autofocus

Some laser cutters are equipped with an autofocus mechanism. This device consists of a plunger and a limit switch that informs the machine of the material thickness when the autofocus operation is complete. During the autofocus sequence, the laser cutter moves the lens assembly to a predetermined location and then slowly raises the bed until the plunger is depressed. It then lowers the bed to relax the plunger and begins the cutting or engraving process. Although this tool is a useful addition to any laser cutter and reduces the risk of improperly focusing the laser, there is one big caveat: if you are using material that has already been cut, there is a chance that the machine will try to autofocus in an area that does not contain any material. This will result in improper focusing, consequently resulting in a poor cut, a potential for the lens assembly crashing into the uncut material, or in the worst case, the plunger becoming stuck in the honeycomb bed. Autofocusing the laser is thus only recommended for new material or if there is no doubt that the plunger will miss the material.

Job Type

Because the laser cutter is capable of both cutting and engraving, the driver can be configured to do one or the other, or both. This is a very useful setting if you need to isolate certain aspects of your design or just reduce the risk of unwanted operations. A good example of this is if you have already cut out materials but need to engrave a pattern onto their surface, you can print your design, containing both raster and vector elements, and deselect the option to cut vector. The cutter would then only engrave the surface, rather then cut and engrave.

Color-Mapping

Often times, designs require more than one type of laser operation. There might be elements that require more or less power, have different fill types, and so on. The color-mapping tool is designed to assign different laser configurations to different line colors from your design. The result is a limitless list of laser configurations for the different line colors. You can also use this as a means to simplify machine operations. Rather than have the machine operator choose a preset for his material, you can have a preconfigured list of color-coded presets. The operator then changes the design's line color to match the material and print as normal. The driver makes the necessary adjustments for the material type based on line color.

Engraving Settings

Power

The engraving power slider adjusts the laser's power, in percentage, while in operation. The power setting directly correlates to the depth of the engraving. Higher power means a deeper engraving. If the laser chars the material while engraving, use a lower power setting and run it multiple times to achieve the desired depth.

Speed

You use the speed slider to adjust the speed at which the lens assembly moves back and forth during operation. High-speed operation also results in lens assembly oscillations, which decreases the precision of the laser beam. If you are engraving detailed objects, such as vector fonts, set the speed low enough so as to not distort the details.

Image Dithering

The image dithering setting specifies the pattern in which contrast is created during engraving. Each manufacturer implements different methods for replicating an image's contrast. Patterns such as halftone use different sized dots to gain contrast. This method works well for structured images such as clipart. For images with less structured contrast, like photographs, use dithering that employs a more random pattern of dots to create contrast. This method produces images that have a more natural look.

Engrave Direction

As the laser engraves, it can produce a considerable amount of smoke and flame. Because of this, it is better to engrave from the bottom up so that the discolored material is removed during engraving. The result is a significantly cleaner engraving.

Resolution

The resolutionn setting specifies how many dots per inch the laser will produce. Lower settings will produce faint images because the laser spends less time in one area. These low settings are designed for quick test runs to check position and orientation. The most common resolutions are between 400 and 600, which create crisp engravings.

Remember, the print driver will not improve the quality of a low-resolution image. If you are printing images with high resolutions, ensure that they are compatible with the DPI setting.

Cut Settings

Frequency

It might appear that the laser is producing a continuous beam of light. In fact, the beam is pulsing on and off many times per second. The frequency slider is used to adjust the frequency at which the laser is pulsed and varies depending on material type. Plastics tend to use higher frequencies, whereas natural materials like wood, paper, and leather use lower frequencies. Use your machine's presets and adjust accordingly until the desired cut is achieved.

Power

With the power slider, you can adjust the laser's power, in percentage, while in operation. You can adjust as necessary to achieve the optimal cut. Typically, more power results in more heat generated at the cutting point and directly relates to the thickness and density of the material.

Speed

The speed slider dictates how fast the lens assembly moves while cutting and will vary depending on material type and thickness. In general, thicker, denser materials require lower speeds, whereas thin and less dense materials require higher settings. The speed also contributes to how much speed is generated in a specific area and is a helpful adjustment to modify when creating new presets.

Project: Creating Presets for Your Machine

Every manufacturer of laser cutting machines produces a series of recommended settings (Figure 6-12) for cutting various types of material. This is a very useful document, and you should keep it in close proximity to the machine as reference. In the event that the document does not cover a material you would like to cut or engrave, you will have to determine the settings by using the good ol' guess-and-check method.

Step 1

Begin by creating a small 1 in × 1 in square in one of the corners of the build area in your *defaultTemplate.dxf*. This square will serve as one of many test cut pieces that will help you to determine the optimal settings for your laser cutter. Open up your machine's print driver and configure it with a material type that is close in density and thickness to the material you intend to cut.

Figure 6-12. *A laser preset.*

Step 2

Turn on your laser engraver, position your material on the bed, and focus the laser. When the machine setup is complete, send the drawing to the machine, turn on the ventilation, and begin cutting. Note any problem signs as the machine cuts and record them on a piece of paper. These could include excessive smoking, scorching, melting, and so on.

Step 3

Turn off the ventilation and examine the cut. Record the laser's power and speed settings using a marker on the test piece for future reference. Adjust the power, speed, and frequency settings to achieve an optimal cut. Make small adjustments to these settings, and repeat the cutting process on new 1 in × 1 in squares. Each time writing the settings on the piece. This process usually takes a few tries before you reach a set of settings that is acceptable.

Step 4

After you have found the proper settings, open the engraver's print driver and create a new material configuration. Label the configuration file with a name that describes the material type and thickness (Figure 6-13). This will come in handy as your database of configuration files grows.

Figure 6-13. *Write the print settings on the square after the machine finishes. The test pieces can pile up quickly, so this will help you more easily identify the best configuration.*

Machine Mechanics and Operation

Laser cutters cut and engrave material by directing a high-power laser beam through a series of mirrors and lenses that are precisely positioned by the x- and y-axis drive assembly. The focused light is then directed along the tool path generated by the computer's print driver, engraving the surface or cutting through the material. In reality, the laser cutter is quite a simple piece of machinery. With a little know-how, this complex and expensive machine can be operated and maintained in any Makerspace.

> *Laser cutters contain multiple safety switches that disable the high-power laser if tripped. This comes in handy because you can test run the machine with the lid open and the laser will not power on. Make sure that your device has this feature before attempting this.*

The laser cutter is enclosed in a steel box that both supports the mechanical components as well as protects the machine's operator from exposure to the harmful laser. Safety mechanisms are systematically placed around all of the access openings and disable the laser module if tripped. Much like a security system in a house, these switches prevent someone from accidentally opening the protective lid while in operation and being exposed to high-power ultraviolet (UV) laser light. In addition, the enclosure directs the air over the material while the machine is in operation. This negative air flow is generated by the attached ventilation system and is designed to channel exhaust gases away from the workpiece and sensitive optics.

Laser cutters also feature a transparent glass window held in place by a sheet steel lid. This lid is designed to isolate the machine operator from potentially harmful reflected laser light and the flames and fumes encountered while in operation. Although glass provides good UV protection, it is best not to stare into the machine while in operation. Ideally, the operator should wear laser safety glasses to prevent accidental exposure. Keep the enclosure clean by periodically wiping down surfaces with a clean cotton cloth, and remove any stuck debris with an approved cleaner or isopropyl alcohol.

The Axis Mechanics

Laser cutters utilize a three-axis drive system (Figure 6-14) which controls the direction and focus of the high-power laser beam onto or through material. This system is very similar to that found in a standard ink-jet printer and is one of the reasons why laser cutters are handled as printers within their control software. The x- and y-axis are paired together in a gantry configuration that carries the lens assembly along the computed path. The z-axis is designed primarily as a focusing tool, making it possible to cut or engrave a variety of material.

Figure 6-14. *Axis mechanics.*

XY Carriage

The XY carriage consists of two linear y-axis rails that carry the z-axis rail and lens assembly. This lightweight carriage features a series of mirrors that direct the laser beam from the laser cartridge to the focusing lenses within the lens assembly. Proper maintenance is key for providing smooth carriage movement and is probably the easiest thing to maintain on the machine. Periodically wipe the rails clean with a cotton cloth and apply a small amount of approved bearing grease along their length.

Z-Axis and the Cutting Bed

The cutting bed is the surface on which material is positioned when engraving and cutting. This bed often includes a reference ruler and can be realigned as necessary. When engraving, the bed can be made out of any flat material because the laser never passes through the workpiece. For vector cutting, a steel honeycomb sheet is used to support the material. This permits the laser to pass through the material and limits the amount of backside reflection. You can use a pin block in lieu of a honeycomb sheet. This device features a series of reference holes and corresponding pins that you can position under the material. These pins should be located in positions that are not going to be cut and should support the material in three or more locations.

Material Trap

As the laser cuts, small pieces of material will fall through the honeycomb and collect in the material trap (Figure 6-15). You should clean this trap periodically to prevent the debris from accidentally igniting as the laser passes overhead.

Figure 6-15. *Material trap.*

The Exhaust System

Every laser cutter manufacturer provides specifications for that machine's ventilation requirements. These requirements define how many cubic feet per minute, or CFM, of airflow is required to properly evacuate the gases and debris generated from engraving and cutting. Not only does this exhaust pose a potential health hazard, but it can cause damage to the machine's sensitive optics and mechanics.

The typical exhaust system consists of a blower fan and an array of ductwork (Figure 6-16) that routes the exhaust to the outside of the building. The ducting should be made out of flexible aluminum or galvanized steel because plastic ducting is potentially flammable.

Figure 6-16. *Ventilation systems require a substantial amount of space and generate a lot of noise. Ensure your Makerspace has allocated the necessary resources for this system.*

There are alternatives to conventional exhaust systems that utilize a series of filters and a chamber filled with activated carbon. Using this setup, you can operate the laser cutter without venting outside.

Air Assist

Flare-ups often occur as the laser engraves or cuts the workpiece due to the released gases, which can be combustible. These flames can distort the material and cause soot to build up on the lens assembly. Many manufacturers combat this problem by implementing an air-assist system (Figure 6-17) that injects a constant stream of air at the laser's focal point. This suppresses any flare-ups and helps to divert the sooty gases.

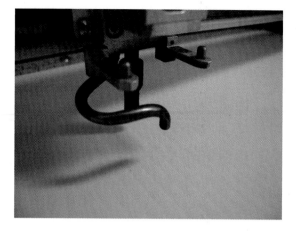

Figure 6-17. *The air-assist tube directs a jet of air at the laser's focal point. This helps to prevent flare-ups and directs soot away from the lens assembly.*

Some manufacturers even offer a combination deep-cut lens and air-assist system that directs the air stream in line with the laser. This not only prevents flare-ups but removes debris from the cut allowing for deeper cuts and better edge quality.

A small compressor is the source of the air, and it should be turned on prior to cutting, so that it can build up a supply, reducing the risk of over-drawing the system. It is important to verify the air-assist's pressure requirement so as not to exceed the limits. A small regulator can be positioned between the compressor and the machine to prevent this problem and provide better pressure stability. Remember, compressors build up water in their tanks as they run. This water should be drained regularly so as to prevent the tank from rusting internally and potentially rupturing.

Engraving

Engraving is the process by which the laser removes a small amount of material from an object's surface, leaving behind a three-dimensional contour. This process can be used to replicate complex images and designs in a method very similar to the commercial ink printing process. Depending on the image type, whether it be vector-based or raster, the laser can repro-

duce this image with high detail using a series of strategically placed dots.

Verify that the material can be engraved by your machine. Some materials might actually reflect the laser light, causing damage to the engraver's mechanics.

Engravable Materials

Engraving metal (Figure 6-18), glass, plastic, and wood is a great way to create permanent markings on the material's surface.

Figure 6-18. *Engraving aluminum.*

Anodized Aluminum

You can engrave anodized aluminum using a CO_2 laser in a method that alters the color of the anodization. When the laser contacts the anodized layer, it changes the base color to white. Black anodization works best because it yields the greatest contrast. Finding the material with the best results will require a little trial and error.

When engraving anodized aluminum, set the machine to operate with high speed and low power settings. Having the power too high will result in the anodization distorting and burning.

Painted Brass

To engrave brass, it must be coated with a laser-compatible enamel. This enamel is removed during engraving leaving behind exposed brass. The laser light does not physically change the surface texture of the brass; thus, the exposed brass maintains its original finish. So, if the brass is polished when it is engraved, it will remain polished after engraving. Brass can be purchased pre-coated with enamel or a specialty spray coatings.

Ensure that the machine is operating with high speed and low power settings. Having the power too high will result in distortion to the enamel and a poor-quality engraving.

Glass

Glass is easily engraved by using a CO_2 laser, and the results are stunning. When the laser contacts the surface of the glass it chips away a small piece, leaving a frost-like mark. The quality and consistency of the engraving depends on the type of glass you're engraving. Sheets of glass have better surface consistency and hardness as compared to bottles, which are less consistent. The resulting inconsistency slightly alters the appearance of the engraving leaving some areas lighter and darker than the others.

Laser Labeling

There are a series of commercially available thermo-reactive coatings such as CerMark that you can apply to the metal's surface prior to engraving (Figure 6-19). These coatings contain compounds that directly react to the laser light and produce a strong black finish that is chemically bonded to the metal. This engraving method is optimal for adding high contrast labels to tools and switch plates.

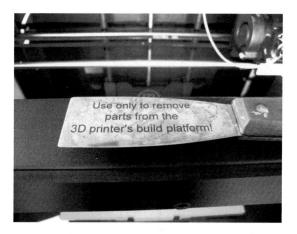

Figure 6-19. *Laser label.*

Plastic

Most plastics engrave nicely due to their consistant density and low melting point. The depth and resolution of the engraving depends on the plastic type. Acrylic and Delrin are good for engraving.

Wood

Woods with consistant grain density, such as bass and maple, work best for engraving. If the wood has irregular density, the depth of the engraving will be inconsistant.

Stamps

The ubiquitous stamp transfers an engraved image onto a surface by using ink or paint. You can use a laser cutter to engrave a stamp. Your image needs to be configured so that the background of the image is engraved away, leaving an elevated surface for the foreground. Also, if there is any text in the image or elements that require specific orientation, you need to "mirror" the image before sending it to the engraver so that the orientation of the stamped image is correct.

Fence

The fence is a polygon that specifies to the engraver the boundary that is to be engraved around your image. This polygon should have a line thickness of 0.001 in, but check with your engraver's manufacturer for exact details.

Shoulders

The shoulder setting in your engraver's print driver defines the addition of a chamfered edge that connects the top of the material to the base of the engraving. A shoulder is not needed if the engraving is thick enough to support itself. If the engraving contains fine details, the addition of an angled shoulder acts to support the position of the engravings' elements. You can adjust this angle to suit your needs.

Cutting

Lasers have the ability to cut material if they can produce enough power at the right frequencies. Most commercial laser cutters possess this ability, and the type and thickness of the material is mainly determined by the laser's wattage. Although some laser cutters have the ability to cut metal, most reside in industry.

To cut material with the laser engraver, it must be compatible with the laser technology and lens assembly. If the material is compatible with the machine, the desired line drawing can be sent via CAD or alternative vector-design software. The print driver is configured to view any line under a certain line thickness, typically <0.001 in, as a vector cut.

Material is cut as the focused beam of light super heats the area around the focal point and the material is melted or vaporized away. This produces a considerable amount of dust and smoke that must be extracted via the ventilation system. If the material is thicker than the laser can fully cut in one pass, multiple passes might be necessary. This method only works for certain types of material; some other types tend to produce an insulating carbon layer around the cut. This layer then absorbs the laser energy preventing it from penetrating deeper.

Cutting Thick Material

You can cut some thick materials on a laser cutter, but it might require multiple passes of the laser (Figure 6-20). This process involves running the machine twice using the same file and material

position. The machine repeats the cutting process and removes the remaining material.

Figure 6-20. *Making multiple passes results in reduced edge quality. If possible, refocus the laser to the depth of the first cut before making the second.*

There are a few problems encountered when cutting thick material. Because the laser is initially focused using the top of the material as reference, making a repeat cut without refocusing the laser results in a loss of power as the laser light diverges. This is generally not a problem for compatible material, but it can produce a considerably larger kerf than a cut made with a single pass.

Alternatively, you can use a lens assembly with a longer focal point, which makes it possible to cut thicker material in one pass. The result is a decrease in the power throughout the focal length because the total energy of the beam is dispersed.

Machine Operation

You can use the following procedure to successfully operate the laser cutter for cutting and engraving compatible material.

Step 1

Turn on the laser cutter and wait for it to complete its startup procedure. When the machine is ready, lift the lid and clear any debris from the bed. Place your material so that it is positioned as close to the origin as possible and verify that it is lying flat against the surface.

Step 2

Focus the laser by using the focus jig and adjust the bed height by selecting Focus on the control panel. Press the up or down arrow buttons until your material just touches the focusing jig. Remove the focusing jig and store appropriately. Exit the focus menu and return the lens assembly to its home position.

Step 3

Start your CAD software on the computer that controls the laser cutter and locate the default template. Confirm that the builds are free of any existing line drawings, open your drawing, and copy and paste it into the template. Position your drawing within the build area so that it is as close to the origin as possible.

If you are using the machine to engrave, verify that the border around your image has been removed and the image is shown as a negative. Because the laser cutter is typically configured to cut white lines, displaying the image as a negative in your CAD application can help to ensure that image will engrave correctly.

If you are using the machine to cut, remove any overlapping lines by using the *OVER-KILL* command (or similar); otherwise, they will result in a repeat cut, which can warp or char your material.

Step 4

Plot your drawing to the laser cutter by pressing Ctrl+P and configuring the print driver for the correct material settings. Click Print to send the job. The name of the drawing should now show up on the machine's LCD. If it does not, power cycle the machine (turn it off and then on again), cancel the print job, and try again.

Step 5

Enable your engraver's laser pointer feature, if available, and lift the lid to prevent the laser from powering on. Press Go to start the simulated engraving/cutting process. Observe the path of the laser and confirm that your material is correctly positioned. If everything checks out, press Stop to discontinue the process and reset the job. When you are ready, lower the lid and turn on the ventilation system with the air-assist, if available. Press Go again to begin cutting or engraving. If a problem occurs, immediately lift the lid to disable the laser and press Stop on the control panel. Pressing Stop is secondary to lifting the lid because it often does not instantly stop the machine; rather, it instructs it to stop at the end of the line.

Step 6

Do not lift the lid until the machine finishes. Wait a short period for your material to cool down and then remove it from the bed. Turn off the ventilation system and pick up any pieces that might be left behind.

Project: Project Box

As your projects become more exciting and complicated, the clutter on your work surface begins to grow. Because electronic projects are integral to most Makerspaces, it is a good idea to have designated soldering areas with complementary support equipment. In the event that your Makerspace is not set in a permanent location, the setup and teardown of equipment can get a bit tedious. This project illustrates a method in which all of the tools necessary to complete an electronics project can be secured into one mobile location. This project box (Figure 6-21) serves as an inexpensive and instantly deployable solder station.

Figure 6-21. *The project box.*

This project requires that safety glasses be worn throughout its entirety.

Materials

You can find the files for this project at Thingiverse.com (*http://bit.ly/16dRNaK*).

Project Materials List		
Material	**Qty**	**Source**
0.2 in × 2 ft × 4 ft plywood sheet	1	Home improvement store
Pegboard hooks	>1	Home improvement store
Small parts box	1	Craft store

Makerspace Tools and Equipment
Laser cutter
Wood glue
Masking tape

Procedure

Step 1

Begin by opening the project file and importing it into the build area of your *defaultTemplate.dxf*. Break the drawing into separate pieces to fit within the limits of your build area and use the *OVERKILL* command to remove any overlapping lines. Send the layout to your laser cutter and repeat the process until all of the pieces have been cut.

Step 2

Begin assembly by applying a small amount of glue to the ends of the center board and insert it into the base board. Apply another bead of glue to the top of the center board and attach the drawer base. Glue to the one side of the base board to the drawer base and attach one of the side boards. Apply glue to one side of the top board and insert it into the corresponding hole on the side.

Use masking tape to hold the sides of the box together. This will help prevent the sides from separating and can help eliminate warping.

Step 5

Apply glue to the ends of the free side and attach the final side board. When all of the boards are seated, apply a bead of glue around the back of the box and attach the back board. Clean up any glue runs using your finger and check that there isn't any glue obstructing the mortises used to secure the front board. Use strips of masking tape to compress the joints together and place the box face down on a table top. Place a heavy object on the back of the box to compress the joints until the glue dries.

Step 7

Position two perforated board hooks into the holes on the center board in a position that can support your soldering iron (Figure 6-22). Install the soldering components into the box and insert a craft bead organizer box into the top slot.

Figure 6-22. *You can configure this box to hold a soldering iron and support equipment. Of course, make sure that the iron has fully cooled before returning it to the box!*

Congratulations! Your handy soldering iron utility box is complete and ready for action.

3D Design and the 3D Printer   7

The availability of 3D printing in Makerspaces has exploded since the first RepRap Darwin was produced in 2007. This is the new wagon headed west. Inspired by its success, the open source community has engineered numerous versions and has made this technology more than affordable for any Makerspace. The 3D printer deposits material layer by layer until the desired computer-generated model is produced. With its almost limitless potential to produce usable objects, it's not hard to see why the 3D printer has become so popular.

Today, there is a wide spectrum of commercially available 3D printers, including the Cube, the MakerBot lineup, Ultimaker, Up, and the list goes on. All of these machines provide a ready-to-go system with which you can create physical objects from virtual models. This chapter aims to dissect the components that make these technologies work and provide you with a better understanding of the ins and outs of 3D printing.

For more information about choosing a commercial 3D printer, check out the Make: Ultimate Guide to 3D Printing (http://makezine.com/volume/make-ultimate-guide-to-3d-printing/), which provides a comprehensive list of reviews. Also, take advantage of the extensive wealth of information on the RepRap Wiki (http://reprap.org)

3D printers have four main requirements: a computer, power, ventilation, and storage. Each 3D printer operates via continued communication with a host computer or attached flash storage. Regardless of the method for delivering the solid model to the 3D printer, there is frequently a computer within close proximity to the printer. Because this computer and 3D printer require an electrical connection and often need to be powered for an extended period of time, a dedicated connection is necessary. Finally, the 3D printer workstation should have access to either active or passive ventilation. This could come in the form of positioning it next to a window or through the use of a local ventilation system. Finally, adequate storage should be available for all of the feedstock and tools required to operate the machine.

Facilitating Design Using 3D Computer-Assisted Design Software

Elevating sketches from two-dimensional profiles to three-dimensional extrusions opens up a world of fabrication possibilities. The parts created within 3D Computer-Assisted Design Software (CAD) provide full-scale virtual representations of your designs and make it possible for you to create prototypes, perform fit-checks, and make tweaks without raising a hammer. Using CAD also reduces the amount of time spent working with physical manufacturing because you can catch potential design flaws by piecing multiple parts together within an assembly. Depending on your software, these assemblies can even reflect motion, interference issues, and help to give a greater understanding of component interaction.

3D CAD utilizes many of the same design tools that are found with 2D, but allows for manipulation in the third dimension. When completed, these three-dimensional parts can be exported as solid-model design files compatible with most of today's manufacturing technology, especially the 3D printer. Although 3D CAD entails a bit of a learning curve, practice makes perfect, and your designs will fly together quicker than you might imagine. Depending on your Makerspace's software style preferences, there are a slew of free and open-source applications that are currently available, with several of them listed in Table 7-1.

Table 7-1. Freely available 3D CAD software

Software Title	Developer	Platform	Link
123D Design	AutoDesk	OS X, Web App, Windows	*http://www. 123dapp.com/design*
Blender	Blender development team	Linux, OS X, Windows	*http://www.blen der.org/*
FreeCAD	FreeCAD development team	Linux, OS X, Windows	*http://source forge.net/apps/ mediawiki/free-cad/*
TinkerCAD	AutoDesk	OS X, WebGL, Windows	*https://tinker cad.com/*

The 3D Environment

If you treat 3D CAD as an extension of your reality, designing in this environment really makes sense. Objects have size, orientation, masses, and features, just as they would in reality, and because so much realism can be integrated within these designs, they can provide a seamless transition into reality. Imagine building with clay on your workbench. The physical orientation of the part and its design features always correspond to a constant plane. This plane, or *work plane*, becomes the platform on which your design is created.

Working with 3D CAD begins with an understanding of the work environment. This includes the base configuration of the software, the different design types, and all of the tools available for designing and creating. With a little patience and a lot of imagination, you will begin to see the amazing potential three-dimensional design has for the production of parts within your Makerspace.

As with any CAD software, design is based around objects drawn within a specified unit system. This option happens to be the first to pop up during new software installation and can generally be found by clicking Tools→Options. Many of today's 3D printers use millimeters as their default unit; you need to take note of the system in which you are drawing. Don't worry if you begin drawing in one system because both your CAD software and the 3D printers control software offer the ability to scale the part to the desired size.

3D CAD utilizes two different types of files: parts and assemblies. Parts are individual objects, and assemblies are complex objects made out of the individual parts. For example, let's consider designing a filament extruder. This mechanism is made out of a series of individually designed parts that can then be virtually combined into an assembly. These assemblies are then used to simulate the interactions of all of the parts and greatly assist with isolating design problems prior to physical construction.

Upon the creation of a new part file, you will be presented with your workspace. The typical workspace is divided into three work planes that define the primary starting points for most sketches. Consider a box sitting on a table. If you look down on the box, you will be looking down on the top work plane. Looking at the front of the box, you see the front work plane, and looking at the right side of the box you see the right work plane. These three work planes are derived from conventional drafting techniques that define a part using the top, right, and front views.

As your part evolves, each surface and the space around the initial three planes can become a new work plane. This makes it possible for you to create components relative to surfaces other than the three primary planes. There are two ways to achieve these secondary work planes. The first is to select a surface and create a new sketch on that surface. Depending on the CAD suite used, this new sketch inherits dependencies for that reference. This means that if the referenced surface changes, the sketch will change, as well. Although these dependencies can be removed, they make alterations and updates considerably easier because you only have to alter one element and all of the elements with dependencies change along with it. The second method is to insert a new work plane relative to one of the primary planes. Using this method, you can create planes that are offset parallel or at a specific angle relative to that primary plane. This method is exceedingly useful if you are aware of the dimensions of the part you are creating.

3D Design Starts with 2D

Nearly every three-dimensional drawing starts with a two-dimensional sketch. The sketch is the "cookie cutter" that defines the design charac-

teristics of what you will extrude into 3D space. Create a new sketch by selecting the work plane of choice; you will be presented with a series of design tools that are identical to those found in 2D CAD. Create the profile that defines your part, and you are ready to extrude.

However, there are few details that need to be addressed prior to extruding a solid model:

1. The profile of your part needs to be fully connected. If you think of this profile as the walls of a cup, if there were holes in the wall, the contents would pour out. For the CAD software to see this as a "fully defined" profile, it needs to be fully connected.

2. You can have design features internal to this profile, such as circles. These features need to be fully defined and will result in a hole being created in your part. This is a useful trick to implement if you know the position of holes prior to the design of your part, and saves the need for creating a new sketch later on.

3. Don't fillet or chamfer the edges of your design until after you extrude the profile. When you enter 3D mode, you will be able to add the fillets and chamfers you want, and it's easier to modify them than to do so in 2D.

After your design is complete and error free, it's time to move into 3D space. This is accomplished by using a series of three-dimensional design tools with which you can extrude, rotate, loft, and sweep your profile as needed. After you have extruded your initial sketch to the desired height, every surface of your new part can become a work plane. This is how you can design complex parts that are not constrained by the initial three work planes. Select the surface you would like to draw on and select the New Sketch tool to create a new sketch on that surface.

3D Drawing Tools

Drawing in 3D is a lot like pottery class: models are shaped, molded, and extruded into virtual volumetric forms that can be physically created using devices like the 3D printer. Consider the way a clay is pushed through a mold. You start out with the profile of the mold and extrude the material until the desired length is achieved. This new structure can then be manipulated into the desired form by using tools that both add and subtract material.

When designing parts for output on a 3D printer, there are a few design tricks you can use to greatly improve the quality of the end product. Most 3D printers cannot print overhangs greater than 45 degrees. This restriction is a direct result of the method in which the printer creates objects by overlapping the extruded filament. If there isn't anything under the filament when extruded, it will sag, creating a spaghetti-like mess (Figure 7-1).

Figure 7-1. *Unless you use support material, a 3D printer cannot print overhangs greater than 45 degrees.*

There are two methods for tackling this problem. First, use support material. This setting tells the G-Code Generator (see "Computer-Aided Manufacturing and G-Code Generation" on page 211) to add small amounts of structure to overhangs. This material can be removed after the print is complete. The second method is to use clever design. When printing objects, like circles that have necessary overhang, modify their design to include an angled cap that does not violate the 45 degree rule.

Alternatively, you can split your model into multiple parts. The overhang constraint imposed by the 3D printer limits objects with top and bottom components. If the model has required components that violate the printer's design constraints, break the model into multiple pieces and assemble it with fasteners to achieve the complex shape.

The following tools are used by most 3D CAD software in the creation of solid models. These tools are designed to create and manipulate models in three-dimensional space, and with a good understanding of their fundamentals, anything is possible.

Chamfer
> You use the chamfer tool to bevel the edge of a part. It works on both convex and concave edges. You can use this tool to provide a symmetrical or asymmetrical bevel to the edge whose lengths and angles can be defined. Not only do chamfers improve the appearance of parts, they are also used to improve the strength of joints by more evenly distributing forces between two faces.

Extrusion
> You use the extrude tool to pull design profiles into three-dimensional space. This is the primary tool used for the initial creation of solid models. Upon selecting the tool, you are prompted to enter a series of parameters that define the extrusion. The first of these is to select the desired profile. As mentioned before, the profile must be complete and without extraneous elements, like floating line segments. The next parameter to define is the direction and length of the extrusion. This feature dictates just how long your extrusion will be and in what direction relative to the work plane.

Extrude Cut
> The extrude cut tool is identical to the extrude, except it removes material rather than adding it. Upon selecting the tool, you are prompted with the same parameters as

those for the extrude tool, except the extrusion must be made in the direction that is internal to an existing part. Imagine this tool as a cookie cutter; you select the shape of your cutout, direct it toward your part, and set the depth of the cut. This tool provides a great way for removing material from a part and creating design features such as channels and holes.

Fillet

You use the fillet tool to make a rounded edge of a part with a specific radius. It works on both convex and concave edges. Not only do fillets improve the appearance of parts, but they are also used to improve the strength of joints by more evenly distributing forces between two faces.

Loft

The loft is used to produce a software-defined connection between two profiles. This connection can either hollow-out an object or make it solid, making complex contours possible. A good example is in the design of a boat hull. The contour can be created by producing a series of profiles that define the primary hull characteristics. Then, a loft can be created from the bow to the stern, filling in the space between the profiles.

Measurement

Measurements and dimensions are utilized both in 2D and 3D design. After the part has been created, you can use the measurement tool to establish the distance between two parts or elements. This tool is handy for verifying the dimensions of your design prior to manufacturing. You can also use it to alter the existing dimensions if the part is properly constrained.

Patterns

You use the pattern tool to create multiple copies of a pre-existing component or body. When you have completed the desired design element, use this tool to select the design element to make copies. This tool can create a linear array or radial pattern and is considerably helpful when designing a part with multiple identical elements, such as spokes for a wheel or the teeth of a gear.

Revolve

You use the revolve tool to rotate a profile around a central axis to any desired degree. Design with this tool is used by creating two design features, the profile and the axis of revolution. Create a new sketch on the desired work plane, draw the profile, and make sure it is complete. Create a new sketch on the same work plane at the location where you want your profile to be revolved and draw a short line. The length is arbitrary: it's only used as a reference. When complete, select the revolve tool and you will be prompted to select a profile and an axis for revolution. Next, set the degrees of revolution, use 360° for a complete revolve.

Revolve Cut

This revolve cut tool is identical to the extrude tool and similar in function to the extrude cut. Follow the same method for creating the profile and axis of revolution used with the revolve tool, except ensure that the profile will be swept through a previously existing model. This tool is handy for creating complex cutouts that follow an arc through a part, such as creating a cavity within a block when designing a mold.

Sweep

The sweep tool provides similar functionality as the loft tool, except the contour of the sweep does not change along its path. To use this tool, you first create a profile and then a path that extends perpendicularly to the surface of the profile. The sweep tool then extrudes the profile along the defined path. You can use it to create a hollow or solid object. Sweeps are often used to represent wire paths through a machine in an assembly or when defining a part's complex bends.

Project: 123D Design for 3D Design

A commonly overlooked accessory for a 3D printer is a holder for the filament spool. This filament is often awkwardly propped against the machine, placed on the floor, or even hung from the ceiling. Consequently, the result is a tangled mess that requires constant attention. This project is designed to provide a quick introduction to the world of 3D modeling through the use of Autodesk's free 123D Design software to design a simple filament holder (Figure 7-2) for your Makerspace's printer.

Figure 7-2. *Filament spool holder.*

The software provides the basic tools necessary to produce and edit solid models. When you start it, you are greeted with an empty workspace and a default unit system set to millimeters. Across the top of the window are the editing tools followed by the viewport selection widget. One of the perks of using 123D Design is its user-supported online database of models that you can import through the parts library. When parts are saved, you have the option to save them locally or upload them directly to your account with Autodesk. From there, you can make your drawings public and share them with the world.

Because 123D Design does not feature a visible design feature history or design tree like that found in Inventor and Sol-

idworks, it is important to frequently save multiple revisions of your work. This will allow you to go back in time and make the necessary adjustments rather than overwhelming the undo button.

Materials

You can find the files for this project at Thingiverse.com (*http://bit.ly/13GKOge*).

Materials List		
Item	**Quantity**	**Source**
1/4 in × 36 in hardwood dowel	3	Home improvement store
8 mm × 22 mm ball bearings	2	Sporting goods store
M8-1.25 × 150 mm threaded rod	1	Home improvement store
M8-1.25 nut	2	Home improvement store
M8 washer	4	Home improvement store

Makerspace Tools and Equipment
3D Printer
Band saw
Sandpaper
Epoxy
Computer with CAD

Designing the Spool Structure
Step 1

Begin by selecting the Top viewport by clicking the top of the viewport widget. This reorients your screen so that you are looking down on the initial sketch plane. Highlight the Sketch toolbox and select the Polygon tool. You will use this tool to produce the primary structure of the spool (Figure 7-3). After selecting the polygon tool, click the sketch plane and place the center point at the plane's origin. Move your cursor to the right until the measurement reads 25 mm and set the number of sides to 6. Click to complete the drawing. Alternatively, you can enter 25 mm into the measurement box and press enter. Press Esc to exit the tool and save your work.

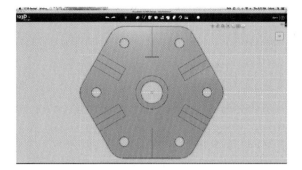

Figure 7-3. *The spool structure.*

Step 2

Select the area within the polygon and then, from the Sketch toolbox, select the Circle tool. Move the mouse to the origin of the polygon and create a circle with a 15 mm diameter. Again, select the area within the polygon and then, again from the Sketch toolbox, select the Polyline tool. Create an 8 mm long line segment that extends from one point of the polygon inward at 30 degrees relative to the origin and then exit the tool. Select the area within the polygon and create a 6.5 mm wide circle at the end of the line segment and then exit the tool. Select and delete the line segment. The polygon and circles you just created will act as a "cookie cutter" that defines how the object will be extruded. Press Esc to exit the tool and then highlight the area outside of the center circle. This defines the area that is to be extruded by using the Extrude tool. From the Construct toolbox, select the Extrude tool and extrude the polygon shape upward 12 mm. Exit the tool and save your work.

Step 3

Create an array of holes by changing your view to the Home position and then, from the Pattern toolbox, select the Circular Pattern tool. Select the visible wall of the 6.5 mm-wide hole to set the faces to pattern. Change the selection mode to Axis by click-

ing the Circular Pattern tool's axis selection cursor. Select the visible wall of the center circle, change the pattern count to 6, and then press Enter. This should create a pattern of 6 holes around the center circle. Save your work.

Step 4

Change the view back to Top, select the area around the top of the extruded shape, and then select the Circle tool. Selecting the area prior to selecting the tool informs 123D on which plane you want to draw and makes the drawing process less confusing. Create a 22 mm circle at the origin of the polygon and exit the tool. Select the area inside the circle and then select the Extrude tool. Rather than extrude new material, you will be removing material from the body of the extruded polygon. Set the dimension of the extrude to −10 mm, click the drop-down menu next to the measurement text box, and then change the extrude type to Cut. Press Enter and the 22 mm circle should have cut a 10 mm-deep hole extracted from the extruded polygon. Exit the tool and save your work.

Step 5

Change your view to Front and select the Rectangle tool. Upon selecting the tool, a dialog box opens next to your cursor asking you to select a sketch plane. Select the side of the extruded polygon facing you and then, with the "Sketch" toolbox, select the Rectangle tool. Create a 6.5 × 6.5 mm square located at the new origin and exit the tool. Select the area inside the square and then move your cursor over the Settings icon (the gear-shaped graphic). Select the Move function and click the arrow that represents the X-axis. Move the square along the X-axis 3.25 mm and the Y-axis 2.75 mm so that it is centered on the side. Again, use the Move function to rotate the square by dragging the rotation handle and setting the angle 45 degrees. Exit the tool and change your view back to the Home position. Select the area within the circle and make an extrude cut 15 mm into the body of the shape. Use the Circular Pattern tool to pattern the extrude cut 6 times around the object (Figure 7-4). You might have to select all five sides of the rectangular cutout for the operation to complete. Exit the tool and save your work.

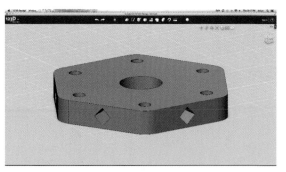

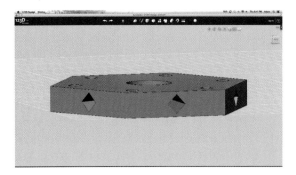

Figure 7-4. *Square-shaped holes can be used in lieu of circles because their 45-degree corners are compatible with a 3D printer's inverted feature constraint.*

Figure 7-5. *Adding a fillet to corners, edges, and junctions not only reduces material and improves appearance, it also increases joint strength.*

Designing the Spool Mount
Step 1

Now, it's time to work on the mount (Figure 7-6). Start by creating a 20 mm hexagon located at the origin by using the Polygon tool. Select the area inside the polygon and draw an 8 mm diameter circle. Using the Polyline tool, draw a line starting at the origin of the 8 mm circle and connect it to the edge of the circle along the x-axis so that it snaps to the grid. Continue the polyline upward until it snaps to the edge of the polygon. The resulting line should be 17.691 mm long. Repeat this process for the other side of the circle and then exit the tool. This makes a channel that facilitates loading of the spool into the holder. Select the area within the polygon and then, from the Sketch toolbox, select the Trim tool. Remove the unnecessary lines by clicking them when they turn red. Extrude the resulting shape 12 mm and save your work.

Step 6

From the Modify toolbox, select the Fillet tool and select each vertical edge of the polygon extrusion. Set the fillet radius to 10 mm and press Enter. This rounds the edges of the polygon and speeds up the build process by removing unnecessary material (Figure 7-5). Save your work, open the 123D Design menu, and then click Export to save the drawing as a 3D printer-compatible *.stl* file. The part is now ready for printing. Open the part in your 3D printer's control software and print away!

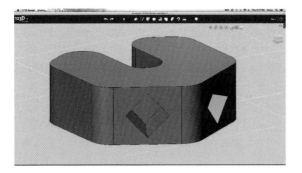

Figure 7-6. *The spool mount.*

Step 2

Change your view to the Home position and create a 6.5 × 6.5 mm square on the lower-left side of the polygon shape. Use the Move tool to move the square to the center of the side and rotate it 45 degrees. Extrude cut the square 10 mm into the body of the shape. Use the Circular Pattern tool to copy the extrude cut to the opposite side. You can do this by selecting 5 surfaces of the extrude cut, setting the number of instances to 2, and defining the 8 mm hole as the axis. Change the pattern type to Angle and change the angle to 60 degrees by dragging the indicator arrow until the angle dialog box opens. Exit the tool and save your work.

Using the Array tool, you can repeat an element in a linear pattern or around an axis (Figure 7-7). Make sure you select every element of the feature you want to pattern before finishing the tool.

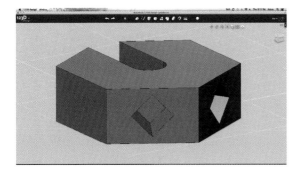

Figure 7-7. *The Array tool.*

Step 3

Select the Fillet tool and fillet the vertical edges of the polygon with a 5 mm radius. When complete, save your work, export the file as an *.STL*, and print away!

Designing the Spool Feet
Step 1

The final step in this project is to design the spool holder's feet (Figure 7-8). Begin by creating two 12 × 25 mm rectangles located at the origin and perpendicular to each other. Extrude the shape up 12 mm. Create a 6.5 × 6.5 mm square located at the center of the end of one side, rotate it 45 degrees, and then extrude cut it 13 mm into the shape. Repeat this process for the remaining side and then save your work.

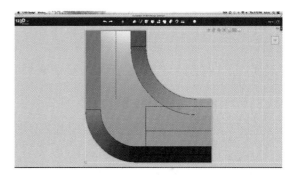

Figure 7-8. *The spool holder feet.*

Step 2

Select the Fillet tool and fillet the vertical edge at the origin and the crux of the shape with a 10 mm radius. From the "Modify" toolbox, select the Chamfer tool and apply a 3 mm chamfer to the top and bottom edges. When complete, save your work, export the file as an *.STL*, and print away!

Making the Filament Spool and Holder
Step 1

To construct the filament holder, start by cutting all 15 × 4 in, 4 × 9 in and 2 × 6 in pieces of the 1/4 in dowel by using the band saw. Gently sand the edges of the cut dowels to prevent splinters and to assist with insertion into the plastic parts. Assemble the spool by mixing up a small quantity of epoxy and gluing 5 × 4 in dowels between the spool holders. Ensure that the bearing mounts are facing out. When these are dry, glue the remaining 10 × 4 in dowels onto the ends of the holder. Push one bearing into each bearing holder (Figure 7-9).

Figure 7-9. *You can use a vise to press the bearings into their seats.*

Step 2

Assemble the spool mount by mixing up another batch of epoxy and gluing the 4 × 9 in dowels into the spool mounts. Glue one foot to the end of each 9 in dowel and connect together using the 2 × 6 in dowels.

Step 3

Slide the M8 threaded rod through the two bearings and lightly secure in place with a nut on each side of the spool. Place a washer over each end of the threaded rod and place the spool onto the holder. Add another washer to each end and secure in place with a nut. The spool should now spin freely and is ready for use.

CNC Fundamentals

Computer controlled machines have been around since the mid-1900s and were designed to replace the human operator, providing greater consistency and the ability to work without

break. The concept of Computer Numerical Control, or CNC, has made its way into the Makerspace environment in the form of CNC mills, lathes, and most recently, the laser cutter and 3D printer.

The CNC process starts with a two- or three-dimensional drawing made in CAD. This drawing is then sent to a processor, or *G-Code Generator*, where it is deconstructed into a series of machine movements based on the size of the tool and capability of the machine. The resulting code, or *G-Code*, is then systematically fed into the CNC control system, which interprets each line of code and moves the axis and tools accordingly. It really is a fantastic process to witness and with a little understanding of each step of this process, your Makerspace will be better equipped to operate and maintain its most sophisticated equipment.

CNC Hardware

CNC machines are identical to their hand-driven counterparts but utilize indexable motors in lieu of hand wheels. Machines such as the 3D printer have three axes, which enables the movement of the extruder to any position in its three dimensional build area. These machines can be set up in three configurations to achieve three-dimensional movement.

3D Printer Axis Configuration
1. Configuration 1
   a. Extruder moves on x- and y-axis
   b. Build platform on x-axis
2. Configuration 2
   a. Extruder moves on x- and y-axis
   b. Build platform moves on y-axis
3. Configuration 3
   a. Extruder moves on z-axis
   b. Build platform moves on x- and y-axis

Each axis consists of a motor (whether it be a stepper or servo) that is directly connected to a drive mechanism. This drive mechanism translates the rotation of the motor into the linear motion required by the axis. As the motor turns the drive mechanism, the axis carriage is moved to the specified linear position. The position of the carriage is determined by using end-stop devices that detect the minimum or maximum linear position.

Stepper Motors

The stepper motor is a brushless electric motor that produces controlled output rotation from a series of electromagnetic pulses. Contrary to a brushed DC electric motor, which produces continuous rotation when connected to a power source, the stepper motor requires a driver circuit that controls the position of the rotor as it rotates.

Each motor is made up of a multiphased stator containing a series of toothed coils and a multi-toothed rotor (Figure 7-10). Motion occurs when the controller sends a pulse to one of the coils, which aligns the poles of the coil to that of the rotor. The typical stepper motor requires 200 steps to complete a rotation, which works out to 1.8 degrees of rotation per step. Because a stepper motor can only execute one step at a time and each step is at the same angle, feedback is not required, resulting in relatively simple drive system. The dimensions of each motor are standardized by the National Electrical Manufacturers Association (NEMA), and are illustrated in Figure 7-11. Table 7-2 lists the standard dimensions.

Figure 7-10. *A stepper motor has a multitoothed rotor that is pulled in line with a series of coils.*

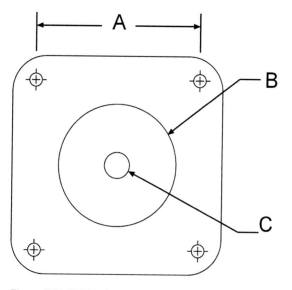

Figure 7-11. *NEMA stepper motor.*

Table 7-2. NEMA dimension chart

Frame Size	A (mm/in)	B (mm/in)	C (mm/in)
8	16.0/0.630	15.0/0.591	4.0/0.157
11	23.0/0.906	22.0/0.866	5.0/0.197
14	26.0/1.024	22.0/0.866	5.0/0.197
17	31.0/1.220	22.0/0.866	5.0/0.197
23	47.14/1.856	38.1/1.50	6.35/0.250
34	69.6/2.74	73.02/2.875	12.7/0.5

The power output of a stepper motor is based on how much torque and speed the motor can produce. Typically, manufacturers present a motor's voltage, current consumption, holding torque, and steps per rotation on a label located on the side of the motor or within the motor's specification sheet. In general, the more power the motor consumes, the stronger it will be. Although more power sounds good at first, it might not be necessary. Choose a motor that is best for the task by determining what kind of drive system it will be turning and what kind of load it will be moving. The motors needed to move the x- and y-axis of a 3D printer don't necessarily need to be the same size as the z-axis motor. Making informed decisions about the type of stepper motor you should use in your project will lead to better performance, less strain on the control system, and lower component costs.

The most common types of stepper motors are 6-lead unipolar and 4-lead bipolar. Unipolar stepper motors contain a center "tap" that bisects each coil so that the motor can be driven without changing the coil's polarity. Bipolar stepper motors contain a single coil per phase and require the controller to reverse the polarity on the coil in order to change direction.

> You can use unipolar stepper motors as bipolar by attaching the ends of each coil to the bipolar driver and disregarding each coil's tap.

The color coding (Tables 7-3 and 7-4) for stepper motors can vary from manufacturer to manufacturer and can be easily determined just by using a multimeter. Configure your multimeter to measure <10 Ohm resistance and create a list of wire colors. Connect the multimeter's leads to any two wires, note the resistance, and continue for all combinations. When you have located pairs of wires with the least resistance, you have found a coil.

Table 7-3. Common unipolar and bipolar stepper wiring

	A1	A Tap	A2	B1	B Tap	B2
Code 1	Black	Yellow	Green	Red	White	Blue
Code 2	Orange	White	Blue	Red	Black	Yellow

Table 7-4. Common bipolar stepper wiring

	A1	A2	B1	B2
Code 1	Black	Orange	Red	Yellow
Code 2	Black	Green	Red	Blue

Drive Mechanisms

Linear drive mechanisms are designed to translate the rotational motion of the motor to the linear motion of the carriage. This is accomplished by directly coupling the motor to the carriage using one of the following methods.

Motor Coupling

The most important aspect in the linear drive mechanism is the method in which the motor is coupled to the system (Figure 7-12). This is accomplished by attaching a pulley or coupler directly to the shaft and securing in place with a set screw. Methods that employ a piece of flexible tubing to couple the motor to the drive should be avoided because they produce a considerable amount of backlash or unintended rotation.

Figure 7-12. *Motor coupling.*

You can make a simple shaft coupler (Figure 7-13) out of a bolt that is a few sizes larger than the shaft diameter. Using a drill press or lathe, bore the center of the bolt to the diameter of the motor's shaft. When this is complete, drill and tap a hole in the head of the bolt for the set screw.

Figure 7-13. *A simple nut/bolt shaft coupler.*

Screw

Screw-drive mechanisms are predominantly used with CNC machines that require a considerable amount of linear stability (Figure 7-14). The mechanism consists of a threaded rod that interfaces with a threaded coupling on the carriages for the axis. When the threaded rod rotates, the coupling is moved along the direction of the threads. This drive method is commonly used for 3D printer's z-axis because less strain is placed on the axis motor when the axis is not in motion.

Figure 7-14. *Screw-drive mechanism.*

Belt

Belt-drive mechanisms consist of a reinforced timing belt that interfaces with a toothed pulley attached to the drive motor (Figure 7-15). As the motor rotates, the belt is fed through the pulley and pulls the carriage to the desired position. The amount of torque transferred to the carriage directly correlates to the diameter of the pulley attached to the motor. Larger pulleys produce less torque and more motion per step, whereas smaller pulleys produce more torque and less motion per step.

Figure 7-15. *Belt-drive mechanism.*

Linear Guides

The linear guide consists of a precision ground rod or rail that supports the axis's carriage as it travels. The carriage couples to the guides by using either bushings or bearings, providing smooth and resistance-free travel.

Rods

Linear rods are made out of rigid metal and provide a cost-effective method for achieving smooth linear motion (Figure 7-16). The cheapest linear rods are made out of ceramic-coated aluminum; more robust rods are made of hardened steel. The carriage is attached to the rods by using round bushings or bearings coupled to a support structure.

Figure 7-16. *Linear rod.*

Figure 7-18. *Linear rail.*

Linear rods can be scavenged from old inkjet printers (Figure 7-17). You can acquire these printers either for very little money or even free from yard sales and thrift stores. Two identical printers will produce enough components to construct a high-quality axis.

Figure 7-17. *Linear rod in an inkjet printer.*

Rails

Linear rails are designed to provide a smooth linear motion and come pre-equipped with mounting features (Figure 7-18). The carriage is attached to the rails using bearings or low-friction slides. Linear rails are often more expensive than rods but provide much greater load capacity.

Bushings

Bushings are made of nylon, PTFE, PTFE-coated aluminum, or bronze due to the material's strength and low friction (Figure 7-19). These devices are less expensive than bearings but provide a good method for achieving smooth axis motion at a lower cost.

Figure 7-19. *Linear bushing.*

Bearings

Bearings provide the smoothest linear motion but are significantly more expensive than bushings (Figure 7-20). Roller bearings consist of an outer roller that supports a series of hardened steel balls and a *race*. Either the roller or race can be secured, allowing the other to spin freely. Linear bearings consist of a series of channels filled with small ball bearings. The balls travel with the motion of the bearing, providing smooth linear motion.

Figure 7-20. *Linear bearing.*

Computer-Aided Manufacturing and G-Code Generation

For a computer-generated solid model to be extruded into reality, it needs a bit of processing. CNC equipment follows a tool path that is generated by Computer-Aided Manufacturing (CAM), software that is configured with all of the operational specifications for your machine. Layer by layer, the software creates a string of *x* and *y* movements and a slew of other commands that control the printer's accessories, including the extruder and heaters.

CAM Software

There are free and open-source CAM packages available that are designed specifically for use with 3D printers (Table 7-5). Although each is unique in their own way, and some are even tailored for one specific type of machine, they all accomplish the same task. After the CAM software is installed and open, you can import solid models into the available build space. This space is represented by a three-dimensional work plane that represents the extent of the build

platform. The model then needs to be properly oriented so that the base of the object is placed onto the build platform. After the build platform is populated with all of the desired parts, the CAM software utilizes its G-Code Generator to produce the code necessary for printing. The resulting G-Code is then sent to the controller via a constant data stream or local storage.

Table 7-5. Freely available CAM software

Software Title	Platform	Link
ReplicatorG	Windows, OS X, Linux	http://www.replicat.org/
RepRap DriverSoftware	Linux, OS X, Windows	http://www.reprap.org/ wiki/DriverSoftware
MakerWare	Linux, OS X, Windows	http://www.maker bot.com/makerware
Repetier-Host	Linux, OS X, Windows	http://www.repetier.com/ download/
Pronterface	Linux, OS X, Windows	https://github.com/ kliment/Printrun

The first thing everyone should print when they get their printer up and running is the beloved *calibration cube*. This 20 × 20 × 10 mm cube is designed to act as a quick means for determining your machine's layer quality and height and dimensional stability.

An alternative to the cube is the *calibration vector* (Figure 7-21). This little device serves as both a means for determining things like layer quality and height as well as enough surface area to accurately determine your axis alignment. Particularly with the z-axis, axis alignment is easy to adjust and dramatically affects build quality. You can find the files on Thingiverse (*http://bit.ly/ 174JLW4*).

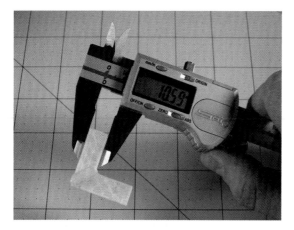

Figure 7-21. *Calibration vector.*

G-Code Generator

The G-Code generator (Table 7-6) is responsible for *slicing* a solid model into layers that direct the 3D printer. These toolchains contain virtually every parameter required to systematically disassemble a solid model, analyze it, and reassemble it into a series of layers.

Table 7-6. Freely available G-Code generators

Software Title	Link
Cura	https://github.com/daid/Cura
SkeinForge	http://fabmetheus.crsndoo.com/wiki/index.php/Skeinforge
RepRap Host Software	http://reprap.org/wiki/Reprap_host_software
Slic3r	http://slic3r.org/
Miracle Grue	http://makerbot.github.com/Miracle-Grue/

Layer Height or Thickness

The height of each layer is dependent both on the extruded filament's thickness as well as the desired amount of detail in the printed object. Adjusting this setting sets the amount of lift the z-axis makes before beginning each layer.

First Layer Height

Because the first layer of the printed object needs to be the most robust, its height is often greater than the subsequent layers. This setting should be adjusted to optimize the first layer adhesion to the build platform and to provide proper support for the rest of the object.

Fill Density or Infill

The fill density setting configures the quantity of material that is deposited between the walls of the object. You might want to lower this when making test prints of a model because it greatly increases build speed. When the model is verified, the infill should be increased to between 30 and 50 percent to produce a sturdy object. For objects that need to withstand substantial forces, the infill can be increased to over 90 percent, which will result in a virtually solid object. Be careful with higher fill densities because parts tend to delaminate as the material shrinks.

Fill or Infill Pattern

The fill pattern settings establish the method on which the extruder fills the space between the object's perimeter. Rectilinear is a good choice of pattern setting, producing a strong, sectioned interior structure.

Print Speed or Feedrate

The print speed determines how quickly the extruder moves over the build platform. It has two primary settings that adjust the speed at which the printer prints and how fast it changes position when not printing. This setting should be adjusted until optimal print results are achieved, and directly affects the flow rate. This setting is dependent on both the strength of the motors and the extruder's ability to consistently deposit material. If the setting is too high, the extruder will not be able to keep up, which will result in thin and choppy layers or the axis steppers will skip steps, offsetting the print.

Skirt or Initial Circling

The skirt produces an initial outline around the perimeter of the object prior to construction. This skirt is used to prime and clear any old material from the extruder prior to printing, which contributes to significantly better print jobs.

Raft

The raft setting enables the addition of an initial removable base layer of material that acts as a buffer between the build platform and the model. The raft helps eliminate problems with nonlevel build platforms and improves adhesion, because the raft is made by depositing large amounts of extruded filament onto the build platform. After the raft has deposited its initial layer, one or more thin layers of material are deposited that both support the object's first layer and allow for raft separation after printing.

Flow Rate

The flow rate directly relates to the print-speed setting. This setting determines how quickly the extruder prints material as it creates each layer. This setting should be adjusted until the desired layer width is achieved.

Width over Thickness

The layer-width over thickness setting describes how wide the layer is as it is extruded. This setting directly correlates to the flow rate and layer height because they determine how much material is deposited. You can adjust this setting after physically measuring the thickness of a single deposited extrusion.

Support

The support feature adds weak woven support structure to any overhanging elements and can be easily removed after the build is complete. This added material helps to prevent problems that occur when overhangs exceed the 45-degree design constraint. Printing overhangs without support results in the extruded filament sagging due to the lack of material upon which to build.

G-Code

G-Code is a numeric control language that is used to control CNC machinery. For 3D printing, G-Code consists of two command types: G-codes

and M-codes. G-Codes are preparatory commands that instruct the machine where to go and how to get there (Table 7-7). M-Codes are miscellaneous commands that directly affect the controller and its interfaces (Table 7-8). These commands are assembled into a file that is streamed to the controller during operation.

G-Codes

G-Codes are commands that control the mechanical motion and operation of the machine. These codes contain commands that move the axis at specific print speeds to the desired locations, rotate extruders, and find endstops.

Table 7-7. Useful G-Codes

Command	Function	Example	Notes
G0	Linear movement at rapid feedrate	*G0 X42*	Example will rapidly move × +42 mm
G1	Linear movement at previous or specified feedrate	*G1 X42 F1000*	Example will move × +42 mm at X units/s
G90	Sets absolute coordinates with respect to the origin	*G90*	
G91	Sets coordinates relative to last position	*G91*	
G92	Sets current position as absolute zero	*G92 X50 Y50*	

M-Codes

M-Codes are commands that control the machine's software. These codes contain commands that start and stop the program, set the steps/unit, and many other auxiliary functions. Some M-Codes are modal, which means they retain their setting throughout the program, whereas others complete as soon as the next line of G-Code is read.

Table 7-8. Useful M-Codes

Command	Function
M00	Pauses/stops the program and waits for user input
M01	Pauses/stops the program only if there is user input
M02	Indicates the end of the program

Codes can be manually sent to the controller through a serial terminal program. The easiest method for doing this is to use the Arduino IDE and open up the Serial Monitor. Set the Line Ending to Newline and the baud rate to match the controller, and you should see startup text appear in the window. To jog your x-axis +5.0 mm, simply set the controller to relative positioning by typing G91 in the text bar and pressing Enter. You should get an OK response indicating the controller understood the command. Then type G0 X5 and press Enter; the axis should move 5 mm from its current position, and you should get an OK response from the controller.

The Controller

The CNC controller interprets a data stream of computer-generated paths into the motion of the machine's axis while simultaneously monitoring and controlling a fleet of support hardware. The typical 3D printer control system is capable of driving 4 steppers, two heaters, and a series of sensors that determine both temperature and relative position.

There are quite a few available control systems, such as TinyG (Figure 7-22), that are designed to support 3D printing. Each of these systems takes a new approach at providing the most amount of control for the least amount of money. Table 7-9 presents just a few of the more popular systems.

Figure 7-22. *The GRBLShield, in addition to an Arduino running one of many 3D printer-compatible firmwares, provides an easy to use 3-stepper driver that can quickly become the basis for your 3D printer's control system (https://www.synthetos.com/).*

Table 7-9. *3D printer control systems*

Controller	Type	Processor	Driver	Link
RAMPS	Arduino shield	Mega		*http://bit.ly/ 16xj23z*
TinyG	Stand-alone	ATxmega192	DRV8811	*http://bit.ly/ 14E4A6V*
GRBLShield	Arduino shield	ATmega328	DRV8811	*http://bit.ly/ 16xj3EW*
Gen7	Stand-alone	ATmega644P or ATmega1284P	A4988	*http://bit.ly/ 12qfAai*
MightyBoard	Stand-alone	ATmega1280	A4928	*http://bit.ly/ 1bU94LN*
Printrboard	Stand-alone	AT90USB1286	A4928	*http://bit.ly/ 14XltlP*
Sanguinololu	Stand-alone	ATmega1284P	A4988	*http://bit.ly/ 14k98TF*

Controller Operations

Controllers operate by reading an incoming string of commands and reacting accordingly. These commands are designed to either control the motion of the axial components or set parameters that affect the way the machine and control program operate. Each controller implements a series of features and algorithms that are designed to optimize the motion of the axis in response to each G-Code command.

The following sections describe some of the more common terminology encountered when working with G-code.

Jogging

Jogging refers to the actual motion of the axis and is used when manually positioning the axis by using the control software or by feeding the controller individual lines of G-Code. The distance and print speed determine the quantity of motion that is achieved.

Ramping, Acceleration

Ramping and acceleration refer to motion methods that are designed to optimize the speed at which the tool travels along its path. Because each axis generates a certain amount of inertia as it moves, implementing ramping and acceleration helps to alleviate problems encountered when hard stopping or starting an axis while in motion.

Jerk

Jerk is a derivative of acceleration and refers to the quantity of change in pressure due to variances in the acceleration. This affects the position of the tool as it travels along the path, and you can adjust it to change the quality of the extrusion during positive and negative acceleration.

Stepper Drivers

Axis motion starts with a series of electrical pulses generated by a stepper driver. This device functions as the interface between the low-power controller, whether it be a computer or microcontroller, and the high-power motors. Each driver issues a series of electrical pulses that pull the rotor in a clockwise or counterclockwise rotation, or hold position. This requires a great deal of power to achieve and is one of the reasons most stepper drivers are actively cooled during operation.

Most stepper drivers consist of an dedicated integrated circuit (IC) that is designed specifically for driving stepper motors and other support electronics that promote stable operation. This includes a pair of current-sensing resistors with which the driver can sense how much current is flowing into the stepper, a potentiometer to adjust the allowed current output, and often a pair of microstepping jumpers. The driver is controlled over three primary control lines: enable, step, and direction. The enable pin is used to activate power to the driver's motor outputs. This makes it possible for the driver to hold the motors in position and prevents unintended axis movement. The step input triggers the driver to move the rotor one step and can be triggered at an exceedingly high rate. Some stepper drivers can even be controlled with a modulated frequency. The direction pin sets the direction, whether clockwise or counterclockwise, and the rotor is stepped and controlled with a HIGH or LOW condition on the input line.

You can adjust each driver to increase or decrease the quantity of current made available to the motor. This adjustment is very important to make; having too much current can result in choppy motion and can overheat the motor, whereas too little current might result in missed steps or motor stall. To adjust the current, locate the potentiometer on the stepper driver and set it to the lowest position. Send a command to jog that axis at a low print speed by using the software control panel or during a print if your printer has the capability. While sending the jog command, gradually increase the current until the axis begins to move freely, without hesitation. Stop jogging and the current level should be set. If you notice the axis stalling during normal operations, gradually add more current until the problem is resolved. Remember, it is better to have less rather than more current.

The stepper driver can operate on one of three primary modes of operation: full-stepping, half-stepping, and microstepping.

Full-Step Mode

Full-step mode is motion that is achieved when only one phase is activated at a time (Figure 7-23). This pulls the rotor in line with that phase's coil, and the next coil is activated. If the stepper motor has 200 steps per rotation, each step in full-step mode would result in the rotor turning 1.8 degrees. The benefit of full-step mode is that the stepper motor functions with its maximum rated torque, while sacrificing resolution.

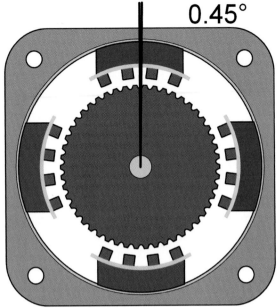

Figure 7-24. *Half-step mode.*

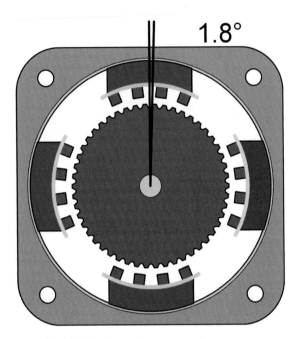

Figure 7-23. *Full-step mode.*

Half-Step Mode

Half-step mode is motion that is achieved when the coils are energized by alternating between one phase and both phases (Figure 7-24). The result is a doubling of the steps per rotation. So, if the stepper motor has 200 steps per rotation in full-step mode, it would have 400 steps per rotation and each step would result in the rotor turning 0.9 degrees. The drawback of half-step mode is the torque fluctuation between one or both phases being energized. This fluctuation is accommodated for, depending on the driver.

Microstepping Mode

Microstepping mode is the most complex of the three modes (Figure 7-25) and is the one enabled on most 3D printers. It requires the driver to servo the current sent to alternating phases with respect to a mathematical function. This makes it possible for the stepper motor to step with resolutions at a fraction of a step resulting in a significant increase in steps per rotation. Each microstepping divisor determines how much the step is divided, which can be as great as 256. This would result in a stepper requiring 51,200 steps per rotation! The drawback of microstepping mode is a loss in accuracy between steps as the divisor increases and an increase in component cost.

0.225°

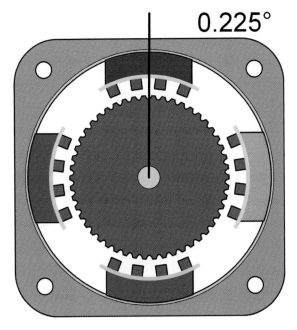

Figure 7-25. *Microstep mode.*

Endstops

Endstop sensors are used by CNC equipment to maintain repeatability and to prevent travel past the mechanical limits of the axis. These sensors detect either the maximum or minimum axis position and relay this information back to the controller in the form of a HIGH or LOW condition. There are four main types of endstops that can be used and each have unique strengths. Regardless of type, the endstop is a critical component for maintaining peak machine performance and product quality.

Hall Effect

Hall-effect endstops utilize a latching hall-effect sensor that detects the presence and strength of a magnetic field (Figure 7-26). These devices consist of two components, a magnet and a sensor, and provide a noncontact method of determining the position of an axis. The endstop produces a digital ON/OFF signal when a magnet attached to the axis comes within a set proximity of the sensor. These devices provide a reliable alternative to mechanical switches, which tend to wear out over time. Care must be taken to prevent accidental switching from nearby magnetic fields, like those produced by motors.

Figure 7-26. *Hall-effect endstop.*

Mechanical Limit Switch

Mechanical limit switches utilize a spring-metal leaf connected to a plunger mechanism that, when depressed, completes the circuit between two contacts (Figure 7-27). There is a fair amount of variation in how the switch is activated and can greatly affect switching accuracy and repeatability. Table 7-10 illustrates the various types and how they are actuated.

Figure 7-27. *Mechanical limit switch.*

Table 7-10. *Limit switch types and activation methods*

Type	Actuation
Pin plunger	In-line
Roller plunger	Slide and cam
Leaf	Slide and cam
Roller leaf	Slide and cam
Lever	Slide and cam
Roller lever	cam

Optical

An optical switch utilizes a light source—typically an infrared LED—and a phototransistor that switches ON and OFF in response to beam exposure (Figure 7-28). Optical switches can be configured to sense when an object breaks a light beam between the transmitter and receiver or transceiver and a reflector, or when light is reflected back when an object intersects the beam. They provide excellent resolution and stability, and when used in the transmitter/receiver configuration, they work great with 3D printers.

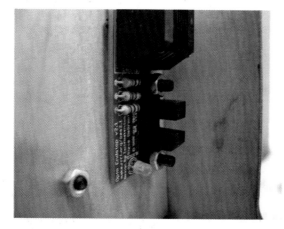

Figure 7-28. *Optical endstop.*

Project: Mechanical Centering, Alignment, and Calibration

Ensuring the proper centering, alignment, and calibration (Figure 7-29) of each axis eliminates many dimensional errors encountered when 3D printing. Axes are centered by determining the machine's travel limits and establishing a center point. This is most important for the xy-axis because it dictates the center of the build platform. Alignment is accomplished by adjusting the relative angle of the axis with regard to the other two axes. Calibration determines the margin of error between the desired steps/unit of travel and the actual.

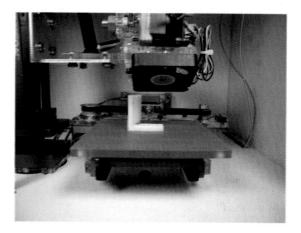

Figure 7-29. *Calibrating a printer.*

Materials

Makerspace Tools and Equipment
3D Printer with Computer
Caliper
Small drill press table vise
Square

Center Each Axis
Step 1

Determining the center point (Figure 7-30) of each axis makes it possible for you to properly place your parts within your control software and maximize your usable build area. Begin by opening your printer's control software and home the axis. This sets the 0-point and serves as the first reference point. Place a piece of paper onto the build platform and lightly tape it in place.

Step 2

Jog the z-axis so that it sits about 1 mm above the build platform and make a small pencil

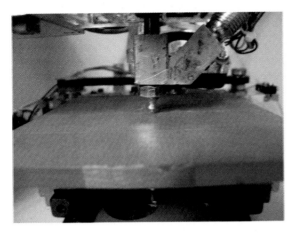

Figure 7-30. *Finding the center point of your build platform ensures that your prints are positioned as intended.*

Figure 7-31. *Use a square to ensure that your second square or straightedge is perpendicular to the build surface.*

mark on the paper, just below the tip of the nozzle. Jog the x-axis away from the endstop until it reaches its maximum travel and then make another mark. Jog the y-axis away from the endstop until it reaches its maximum travel and make the final mark. Repeat this process for the z-axis by using a caliper or ruler to measure the vertical travel.

Step 3

Use a ruler or caliper to measure the length of each line. Divide this length in half to determine the midpoint and then enter the values into the control software configuration.

Determining Alignment

Step 1

Ensure that your printer is set on a flat and level surface to begin alignment. The worktable will function as the starting reference point. From there, you can properly align each axis and determine how level the build platform is. Secure the square into the jaws of the vise and reference its trueness to a second square that's positioned on the tabletop (Figure 7-31). This ensures that the square secured in the vise will serve as an accurate reference.

Step 2

Attach an indicator rod to the z-axis by which to measure offset and lower the axis to its lowest position (Figure 7-32). Position the clamped square in line with the z-axis extension and slowly raise the z-axis. As the z-axis travels, note any change in the distance of the z-axis extension relative to the square. Any deviation will indicate a misaligned z-axis, and you should make adjustments accordingly. Test and retest both the x and y side of the z-axis to verify alignment.

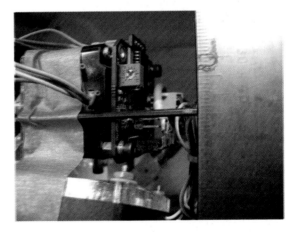

Figure 7-32. *The alignment of each axis is determined by checking the change in alignment over a given length relative to a fixed point. This point will serve as the reference for each axis and will ultimately determine the alignment of the entire machine.*

Step 3

The alignment of the x- and y-axis are not dependent on z alignment but do dictate the levelness of the build platform. On machines with build platforms attached to the z-axis, you can make the adjustment by placing a piece of paper between the tip of the nozzle and the top of the platform. Check the gap at all four corners of the platform as well as the middle and make adjustments accordingly. For designs that feature a moving platform, leveling is a bit more tricky. Start by determining the levelness of the x-axis by using a caliper to measure the distance from the top of one x-axis guide to the top of the table. This measurement will serve as the reference and should be used to adjust the other three guide ends. Repeat this step to align the y-axis relative to the table.

Step 4

Lay the square onto the build platform and move the x- and y-axis to their minimum positions. Power on the machine and use the control software to lower the extruder until the nozzle just touches the edge of the square. Carefully jog the platform or extruder down the length of the square so that it remains in line with the extruder. This aligns the square to the travel of the x-axis and serves as reference for the y.

Step 5

Without moving the square, position the nozzle at the end of the square and jog the extruder or platform down the length of the y-axis. Adjust the guide rails until there is no longer any deviation from the square. Repeat steps 3–5 until the x- and y-axis are properly aligned.

Determining Calibration

Step 1

Power up the 3D printer and open the control software or terminal. If you are using the terminal, set the machine to relative positioning by using the G91 command. The calibration of each axis is determined by com-

paring the desired motion to the actual motion of the axis. You then use this number to calculate the actual steps/unit and is fed into the control software for proper G-Code processing or into the controller's firmware for proper axial control.

Step 2

Center each axis and lock the motors. Carefully secure the caliper in the jaws of the vise so that the dial carriage can slide freely and the depth gauge extends along the axis you are calibrating. Extend the depth gauge by sliding the dial carriage to the side until it touches a fixed surface perpendicular to the axis, such as the side of the printer, an endstop, and so on. Rotate the dial so that the indicator points to 0 or press the Zero button on the caliper (Figure 7-33).

Figure 7-33. *Checking calibration.*

Step 3

Unlock the caliper and jog the axis 0.50 in (5.0 mm), closing the caliper. The caliper should then compress and the difference between the desired motion and actual motion determined. Record the actual steps/unit and adjust the control software accordingly. Repeat this procedure for each axis until calibration is complete.

3D Printing

3D printers produce physical objects from virtual models by using a process commonly known as *fused filament fabrication*. This process involves the extrusion of molten plastic filament that is systematically deposited onto a build platform. Layer by layer, the machine outlines and fills each element until the entire object has been replicated. When the object has cooled, it can be separated from the platform and put to use.

The filament extruder deposits plastic by using a toothed, or hobbed, filament feed wheel in combination with a tensioner. As the feed wheel turns, filament is gripped by the teeth and forced into the barrel located at the base of the extruder. A heater heats the barrel to the temperature required for extruding, which varies depending on plastic type. When the plastic has reached extrusion temperature, the feed wheel forces more plastic into the barrel, in turn pushing molten plastic through a nozzle. Nozzle diameters range in size from 0.3 mm to 1.0 mm and mainly dictate the speed and resolution of your prints.

Extruder Mechanics

Extruders are made up of three primary components: the drive mechanism, the feed mechanism, and the heated barrel. These three components work in conjunction to deposit the desired amount of plastic onto the build platform and should be able to do so for hours and hours. This harsh work environment constrains the types of materials used to build the extruder. Especially within the heated barrel where temperatures are high and fluctuate constantly.

The Drive Mechanism

Feeding filament into an extruder requires a significant amount of force. This force is commonly generated by a stepper motor because of its compact size and step precision. Alternative methods use DC motors and encoders to determine print speed and position, but they add complexity to the system. On most compact extruder designs, the drive motor is attached di-

rectly to the feed mechanism via a pair of drive gears. These gears increase the torque delivered to the feed mechanism allowing for a smaller and less powerful stepper motor to be used. The size and ratio of the gears can be calculated based on the desired steps per rotation of the feed wheel and the overall torque applied by the stepper.

You can use the following formula to determine the approximate steps/mm of filament fed into the extruders and it's a good starting point for tuning your printer:

$$\text{Steps/mm} = R \times r / (\pi * D)$$

R	Quantity of steps per revolution at the motor
r	The gear ratio (r_{feed} / r_{motor})
D	Feed wheel diameter

The quantity of microsteps per revolution is determined by multiplying the degrees per step by the microstepping multiplier. As an example, a stepper motor that rotates at 0.9 degrees per step and is driven with a 4x microstepping multiplier will require 1,600 steps per rotation. The gear ratio is relative to the desired steps per millimeter of extrusion. The most common ratio is 39:11, and when paired with a 7 mm feed wheel, produces approximately 258 steps/mm.

Alternatively the formula can be reworked to solve for the gear ratio. If the extruder were to produce 360 steps/mm, had a 7 mm feed wheel, and was configured for 3,200 steps/revolution, the required gear ratio would be approximately 2.5:1. You can then scale this ratio to accommodate appropriately sized gear, yielding an 8-toothed drive and a 20-toothed driven gear.

The Feed Mechanism

Creating a reliable feed wheel and tensioner pair is the most complex component of an extruder. Without good grip and control, the reliability and accuracy of the extruder is compromised resulting in poor-quality builds.

Hobbed Bolt

The quickest method for making a feed wheel is to use the same bolt that is attached to the extruder's driven gear. You can add horizontally cut teeth to the unthreaded portion of the bolt making for a cheap and reliable alternative to a professionally machined feed wheel. These teeth can be added to the bolt in a process known as *hobbing*, which uses a spinning cutting tool to cut parallel teeth into the material as it turns. Rather then using a proper hobb cutting tool, a thread cutting tap can function as a suitable replacement and can reliably hobb bolts when chucked in a drill press or a lathe.

Hobbed bolts and machined feed wheels are prone to clogging (Figure 7-34) when the filament jams. An easy method for removing stuck filament is to use a wire brush and gently brush along the teeth. You can remove stubborn material with a razor, but take care not to damage the teeth.

Figure 7-35. *Machined feed wheels provide better consistency but at a higher price.*

Tensioners

Working in conjunction with the feed wheel, the tensioner is designed to push the filament into the feed wheel's teeth, placing as little resistance on the feed wheel as possible (Figure 7-36). The most optimal mechanism for the tension consists of an adjustable assembly that forces a bearing against the filament. The more passive approach consists of an adjustable piece of nylon or polyoxymethylene, due to its strength and low friction coefficient.

Figure 7-34. *Teeth clogged with filament.*

Machined Feed Wheel

A better alternative to the hobbed bolt is the machined feed wheel (Figure 7-35). This device functions in the same manner as the bolt, but because it is not cut by hand, it features much higher diameter and thread consistency. Depending on the extruder design, the machine feed wheel can be directly attached to the drive motor or even mounted onto the driven gear's bolt, providing more torque with a smaller motor.

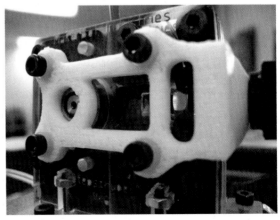

Figure 7-36. *The tensioner ensures that the filament is grabbed by the feed wheel. If the tensioner is too loose, the filament will slip. If it is too tight, the filament may jam and the stepper will skip steps.*

The Heated-Barrel Assembly

The heated-barrel assembly is designed to accept filament from the feed mechanism and elevate it to a temperature high enough to accommodate extrusion. Reaching this temperature involves a combination of components that isolate the hot and cold sides of the extruder, allow for continuous material flow, generate a substantial amount of heat, and provide monitoring and control. When all of these components function as intended, a thin, consistent extrusion of plastic can be generated for many hours on end.

Barrel

The barrel itself has two roles (Figure 7-37). The first of which is to physically isolate the hot end of the extruder from the rest of the device. This is important because there are many problems that can occur when too much heat rises up the barrel. One of which causes the filament to soften and ultimately jam the feed mechanism. To prevent problems like this, the barrel should be made out of a material that is not only strong, but doesn't thermally conduct. There are three materials that are commonly used to solve this problem: PEEK, PTFE (Figure 7-38), and stainless steel. PEEK and PTFE are rigid polymers that can withstand temperatures up to 250°C and have low friction coefficients. Stainless steel is often used because it provides good structural support and resists corrosion.

The inside of the barrel contains a PTFE tube that separates the barrel's structure from the filament. If the filament were to contact the side of the barrel, it would cool, resulting in a filament jam.

Figure 7-37. Heater barrel.

Figure 7-38. PTFE is used inside of the heater barrel due to its low friction coefficient and high heat tolerance.

Nichrome Heater

Nichrome is a nickel-chromium alloy that is used in heating elements due to its natural resistivity and ability to withstand high temperatures (Figure 7-39). The two most common alloys are nichrome A (80 percent nickel and 20 percent chromium) and nichrome C (60 percent nickel, 16 percent chromium, and 24 percent iron). Nichrome A features a lower resistivity that is more stable over temperature fluctuations, whereas Nichrome C features higher resistivity that varies more with temperature. Each wire thickness delivers a known Ohms/ft(m) that can be used to calculate the length of wire required to make the desired heater (Table 7-11).

Figure 7-39. *Nichrome heater.*

Table 7-11. Nichrome resistance/length at 20°C

Gauge	NiCr A (Ω/m)	NiCr A (Ω/ft)	NiCr C (Ω/m)	NiCr C (Ω/ft)
25	6.64	2.02	6.89	2.1
26	8.38	2.56	8.69	2.65
27	10.61	3.23	11	3.35
28	13.35	4.07	13.84	4.22
29	16.81	5.12	17.43	5.31
30	21.15	6.45	21.93	6.68
31	26.92	8.21	27.92	8.51
32	34.38	10.48	35.65	10.87
33	42.4	12.94	44.01	13.42
34	53.71	16.37	55.7	16.98
35	68.2	20.79	70.72	21.56

Soldering Nichrome

One of the problems faced when using nichrome to make a heater is attaching the heater's power wires to the nichrome (Figure 7-40). Nichrome inherently does not bond to solder using conventional electronics soldering flux. This problem can be overcome by using fluxes designed for soldering stainless steel because they also include chrome. Care must be taken when using these fluxes because they contain a large quantity of hydrochloric acid, which is se-

Heater Calculation

Suppose that you want create a 25-watt heater that runs on 12 VDC. You can calculate the required resistance value by reworking the power and dissipated power calculations to solve for the required resistance:

P = IV

P	Power
I	Current flow
V	Voltage

25 W / 12 VDC = 2.083 A

$P = I^2 \times R$

P	Dissipated power
I	Current flow
R	Resistance

25 W / 2.083 A² = R   R = 5.761 Ohm

Therefore, a 25-watt heater would require 5.761 Ohms of resistance at 12 V. If 32 gauge nichrome A is used, the heater would only need 0.567 ft of wire.

verely caustic and must be removed from the solder joint to prevent corrosion. Take time to review the Material Safety Data Sheet (MSDS) prior to use. Alternatively, you can attach nichrome to the power wires by using crimp connectors. This method involves wrapping a small amount of nichrome around the ends of the power wires and crimping the joint in place with a crimp connector. Although the resulting joint works, it is prone to failure resulting from repeated vibration and flexion.

Figure 7-40. *You can solder nichrome to copper wire using some fluxes that are designed for stainless steel and chrome. Thoroughly clean away the flux after use as it will corrode the wires.*

Figure 7-41. *Cartridge heater.*

Cartridge Heater

Cartridge heaters contain a coil of nichrome wire encapsulated in a sturdy stainless-steel tube (Figure 7-41). The high-temperature power leads extend from one side of the element and are bonded to the inside of the cartridge, providing strain relief and extending the heater's lifespan. Because the cartridge cannot be wrapped around the heater barrel, a secondary structure is needed. This "heater block" attaches to the base of the barrel and often serves as the mount for the nozzle.

It just so happens that the automotive industry is a source of reliable heaters. Diesel engines use a glow plug (Figure 7-42) to pre-heat the cylinder to assist with starting the engine. The Autolite 1104 is the perfect combination of size and power for 3D printing, consuming approximately 25 watts at 5 VDC and costing less than half the price of a conventional element heater. You must take care when using this type of heater because they can reach temperatures high enough to start fires.

Figure 7-42. *Diesel heater.*

Power Resistor

The cheapest and easiest solution for heating the barrel is to use a 3-watt wire-wound resistor (Figure 7-43). These resistors are designed to dissipate 3 watts of power in open air. When embedded in a heater block in the same manner as a cartridge heater, the conductivity of the block makes it possible for the resistor to dissipate more power without failing. These resistors are typically made using a nichrome wire wound around a ceramic or fiberglass core and can handle some pretty intense heat. Typical resistance values are 6.8 Ohms, producing ~21 watts at 12 VDC, and 5.6 Ohms, producing ~26 watts at 12 VDC. These resistors are more prone to damage than their cartridge heater counterparts but cost significantly less.

Figure 7-43. *Power resistor heater.*

Nozzle

Nozzles function by tapering the large diameter filament to a desired extrusion diameter that is typically less than 1 mm (Figure 7-44). With designs that use a threaded barrel and nichrome wire, the nozzle has female threads and is screwed onto the end of the barrel. With designs that use a cartridge or power resistor, the nozzle has male threads and is screwed into the heater block. Most nozzles are made out of brass because of its high thermal conductivity and machinability.

Figure 7-44. *Extruder nozzle.*

Brass acorn nuts and hex-head bolts provide a good starting point for machining a nozzle. The nut or bolt can then be bored and drilled using a drill press or lathe until the desired measurements are achieved.

Temperature Sensing

It is important that the temperature of the molten filament is precisely controlled. Variances of only 1°C directly affect the material's flow and can result in inconsistent builds. To accurately monitor the temperature of the filament, extruders feature a thermal sensor that is connected as close as possible to the nozzle. Thermistors and thermocouples are commonly used here because they provide accurate thermal sensing over a wide temperature range. These sensors then relay the values back to the temperature controller, which then modulates power to the heater as it maintains temperature.

Material Types

There are a variety of material types that 3D printers can extrude. The material comes as a long extruded filament typically 1.5 to 3.0 mm in diameter and is fed directly into the extruder. The type of material used ultimately depends on the end application, and each one has unique advantages. The most common materials used for 3D printing include ABS, PC, PLA, PVA, and FRO.

RepRap.org publishes a long list of material suppliers, including cost per kg/lb. To view this list, go to http://reprap.org/wiki/Printing_Material_Suppliers.

ABS

Full name	Acrylonitrile Butadiene Styrene
Extrude temperature	195°C–245°C
Glass transition	95°C

Decomposition temperature	260°C

Acrylonitrile Butadiene Styrene (ABS) is a low cost thermoplastic plastic that is easily extruded and provides good dimensional stability. Primarily used in the automotive industry, ABS lends itself to 3D printing, producing parts that are rigid but not brittle. The filament is available in a wide variety of colors, even glow-in-the-dark.

Because ABS is derived from petroleum products, it is not biodegradable like PLA and produces unpleasant fumes when heated. These fumes can cause nausea if overexposed. If the Makerspace is going to allow the use of ABS filament, ensure that proper ventilation has been installed or switch to a less-offensive filament.

PLA

Full name	Polylactic acid
Extrude temperature	190° C–230°C
Glass transition	60°C
Decomposition temperature	250°C

Polylactic acid (PLA) provides a welcomed alternative to ABS because it is fully biodegradable and produces less offensive fumes when extruded. This polymer is derived from corn starch and is slightly harder and more brittle than ABS. One of the downsides to PLA is that it is hygroscopic (it absorbs moisture), which can lead to extrusion problems. When the water content is too high, bubbles tend to form in addition to inconsistencies in viscosity. These problems are not permanent and can be prevented by storing unused PLA in a dry box or by removing the moisture in a drying oven.

PLA also has a significantly shorter residence time than ABS. If the PLA sits in the extruder for too long while heated, the color and flow properties will be negatively affected, resulting in a brown hue and bubbles. A good remedy for this problem is to perform a purge of the resident plastic prior to building.

PVA

Full name	Polyvinyl alcohol
Extrude temperature	160°C–170°C
Glass transition	85°C
Decomposition temperature	200°C

Polyvinyl Alcohol (PVA) is a polymer found in the food industry because it is fully biodegradable and dissolves quickly in water. Relatively new to Makerspace 3D printing, PVA has shown its usefulness as support material and for making water-soluble parts. If the PVA is left in the heater for too long, or it is exposed to temperatures above 200°C, the material will jam the nozzle. These jams are very hard to remove and often require nozzle replacement.

PC

Full name	Polycarbonate
Extrude temperature	250°C–280°C
Glass transition	146 C
Decomposition temperature	380°C

Polycarbonate (PC) is a strong and impact resistant thermoplastic that requires relatively high temperatures to extrude. Due to these high extrusion temperatures, you must take care to ensure that the extruder will not fail. PC is also hygroscopic and thus will absorb moisture over time, resulting in the same extrusion problems as with PLA.

Wood

Extrude temperature	180°C–230°C

This relatively new material is comprised of wood shavings and a binding polymer that is compatible with PLA and ABS extruders. You can print it at relatively low temperatures and it does not require a heated bed. The resulting prints look, smell, and have the characteristics of any wooden object.

Frosting

Extrude temperature	20°C
Glass transition	20°C
Decomposition temperature	29°C

Plastic isn't the only thing a 3D printer can extrude. Many Makerspace bakers have turned to edible material extrusion. With the use of a frosting-compatible extrusion system, 3D printers are capable of frosting and decorating delectables.

The Build Platform

The build platform provides the starting point for every 3D printed object. It is designed to provide both level support for the build as well as to a bonding surface that prevents the material from shifting and warping as each layer is deposited. To counteract problems with adhesion and warping, many build platforms use designs and materials tailored specifically for each material type.

Platform Design

Thermal expansion and contraction of a material plagues the world of 3D printing. As the extruder deposits material onto the build platform, its molten state has greater volume than when solidified. This difference in volume results in the material changing shape as it cools, often resulting in the part warping and separating from the build surface. There are three methods to counteract this problem: heat the interior of the 3D printer to a temperature high enough to prevent distortion, heat the build platform, or rapidly cool the filament.

Heating the Environment

This method is the most complex and is often only employed with commercial 3D printers. The entire 3D printer is enclosed in a thermally insulated case that is heated by convection. A fan inside the printer blows air past a temperature-controlled heating element that raises the temperature high enough to prevent the material from warping and separating from the build platform. For DIY printers, this environmental heating can be accomplished by enclosing the entire printer in a thermal insulation.

Heating the Build Platform

The most common method for 3D printing materials such as ABS is to actively heat the build platform. This is accomplished by routing a heating element beneath the platform and monitoring its temperature with a thermistor. Some designs use a nichrome wire to provide heating, but you should avoid these due to the nichrome's potential to generate enough heat to ignite the platform, damaging the machine and potentially the surrounding area. A better alternative to nichrome is to use a circuit board with a network of traces designed to produce enough resistance to function as a heater. Because the build platform does not need to reach exceedingly high temperatures, this design provides safer and more reliable heated surface.

Actively Cooling the Filament

Some materials such as PLA can be directly deposited without the need for heating the build platform. This makes for a much less complicated system because two controls, the surface heater and the temperature monitor, are eliminated. For the material to adhere to the surface and not separate when cooled, the extrusion can be slightly cooled immediately after extrusion. You can accomplish this by attaching ductwork and a small fan that blows cool air directly at the extruder's tip. If the extrusion cooled too much it will prevent the layers from bonding. Conversely, if the extrusion remains too hot, it will contract too much when cooled and separate from the platform.

Build Platform Materials and Coatings

Getting hot extrusion to bond to the build platform is not a trivial task. Some bond best when the platform is heated, others do not. Some require special surface coatings, whereas others are less picky. Choosing the proper surface material is just the first step. It then needs to be fine-tuned to prevent the extrusion from sticking too well or not enough. When the configuration is complete, print jobs can proceed with little separation during printing but still be easy to remove upon completion. Table 7-12 lists several printing materials and the platform materials with which they're compatible.

Table 7-12. Build platform material compatibility

Printed Material	Platform Material	Heated
ABS	Polyimide	YES
	Painter's tape	YES
PC	Polyimide	YES
PLA	Acrylic	NO
	Painter's tape	NO
	Polyimide	YES
	Polycarbonate	NO
PVA	Polyimide	YES

Heated

Heated build platforms can be made from materials with good thermal stability and rigidity, including glass, aluminum, and circuit board. Glass platforms provide a good starting point for a build platform and you can use them with or without polyimide or PET adhesive. Glass withstands high temperatures, is mostly level, and very inexpensive. One of the problems faced when using glass is its relatively low thermal conductivity, which can make temperature adjustment a bit tricky. Attach the thermistor to the top of the bed rather than the bottom when used as a heated platform. This helps to more accurately measure the surface temperature. Glass also has the tendency to crack when unevenly heated. You can avoid this by using an aluminum plate between the glass and the bed heater. Because aluminum is a good conductor of heat, it will spread out any hot spots produced by the heater and provide for a more evenly heated platform. The cracking problem can also be solved by using tempered glass. This glass is heat treated and can withstand temperatures and stresses well beyond that of standard glass.

Printed circuit boards can be used as a heated platform by combining the heater, thermal sensor, and structure into one unit. The circuit board material, typically FR4, has good thermal conductivity, thus you can locate the heating element on the top or bottom of the platform. The surface must be coated with either polyimide or PET adhesive to promote extrusion adhesion. The addition of a thin plate of aluminum to the top of the circuit board provides better heat spreading and surface characteristics.

Aluminum provides an excellent surface for printing. Its high thermal conductivity and structural integrity means that it can accommodate large prints with little separation. As with the glass and circuit board designs, aluminum needs to be coated with either polyimide or PET film to promote adhesion. One of the complexities of using aluminum as a build platform is its electrical conductivity. Applying a heating element directly to the bottom of the plate without insulation will result in an electrical short and thus requires isolation. You can do this by tracing the path of the heating element first with a layer of polyimide tape and then place the element. Or, simply by using a heater that is already electrically insulated.

Many scanners use a sheet of tempered glass for the scanner bed. A quick trip to the thrift store can yield an old scanner, complete with glass bed, for the same price as a sheet of nontempered glass.

Unheated

Plastics like acrylic, ABS, and polycarbonate are suitable for nonheated platforms. These materials provide very good adhesion properties that often result in difficulty separating the printed object from the surface. Coating the surface with painter's tape helps lower adhesion and provides an inexpensive method for maintaining the surface quality.

One of the drawbacks of using plastic beds are inconsistencies in thickness and the tendency of the material to bow. These can be avoided by using material at least 1/4 in thick and frequently checking that the platform is level.

Applying a thin coat of cooking oil to the surface of the platform with a clean rag can help lower adhesion, making separation easier.

Coatings

You can coat build surfaces with adhesives that promote adhesion. These adhesives include painter's tape, PET tape, and polyimide tape. Painter's tape is an excellent surface coating due to its low cost and availability. The wax content in the tape promotes adhesion of most extruded materials both with and without heated platforms. Although painter's tape is less durable than other adhesives, its excellent performance and low cost easily make up the difference.

PET, or polyester, film tape is a relatively new material to be used with heated build platforms. This adhesive exhibits approximately the same durability as polyimide tape and has an operating temperature less than 200°C.

Polyimide film tape is a popular surface coating due to its durability and ability to withstand extreme temperatures. This orange tape bonds well to most surfaces and comes in a wide range of widths, lengths, and thicknesses, making it possible for some platforms to use only a single piece of adhesive. Although polyimide is an excellent coating, it is relatively expensive and over time looses its adhesion qualities.

Project: Bolt Hobbing

The drive wheel is the single most important component in an extruder because it provides the mechanical interface between the drive motor and the filament. If there are distortions or unevenness in its design, the filament will not feed evenly, resulting in poor print quality. There are two methods for producing consistent hobbed bolts (Figure 7-45) that will yield hassle-free printing, and both of them can be carried out by using Makerspace equipment.

Figure 7-45. *A hobbed bolt.*

This project requires that safety glasses be worn throughout its entirety.

Materials

Materials List		
Item	**Quantity**	**Source**
8 × 22 mm ball bearings	2	Sporting goods store
M8-1.25 partially threaded bolt	1	Home improvement store
M8-1.25 nut	2	Home improvement store
M8 washer	2	Home improvement store

Makerspace Tools and Equipment
Drill press
Lathe
Safety glasses
M4-0.7 Tap

Making a Hobbed Bolt Using a Lathe

To hobb a bolt by using a lathe (Figure 7-46), you'll need a special bolt holder. The irony of this process is that unless the bolt holder is made by hand, a 3D printer is required.

Figure 7-46. Hobbing the bolt.

Step 1

Search Thingiverse.com for a bolt holder that is compatible with your lathe and use a 3D printer to print it with 100 percent infill. This ensures that the holder is as strong as possible. Install and secure the tap into the jaws of the chuck so that the threaded portion is left exposed. Position the bolt holder into the lathe's tool holder and tightly secure in place. Position the bolt holder such that the bolt is perpendicular to the tap.

Step 2

Use a permanent marker to mark the desired position of the hobbed section onto the unthreaded portion of the bolt. Place two bearings into the holder and slide the bolt into place. Without powering the lathe, move the carriage so that the bolt is positioned approximately 1 mm from the nontapered end of the tap and verify the alignment. Use washers to adjust the alignment and, when set, secure the bolt in place with a nut followed by another. Tightening the second nut will jam the first in place, preventing it from loosening.

Step 3

Configure the lathe to the appropriate RPM for cutting steel. Apply a layer of cutting oil to the tap and turn on the lathe so that the chuck is spinning toward you, or counterclockwise. Slowly move the carriage along the y-axis until the bolt comes into contact with the tap. Apply just enough force to allow the tap to cut and rotate the bolt. Continue cutting until the cuts just meet, resulting in sharp peaked teeth. Cutting too little will result in teeth that are not sharp enough to grip filament, whereas cutting too deep will weaken the teeth. After cutting is complete, back off the carriage and turn off the lathe. Remove the bolt from the holder and wipe off any filings and remaining cutting oil. The hobbed bolt is now ready for action. Install it into your extruder and start printing!

Making a Hobbed Bolt Using a Drillpress

A simpler yet less precise method for producing a hobbed bolt requires only a drill press and a table vise (Figure 7-47).

Figure 7-47. Hobbing with a drill press.

Step 1

Begin by marking the position of the desired hobbed section onto the unthreaded portion of the bolt and slide on two bearings. Position the assembly into the jaws of the table vise so that the bearings sit about an inch away from each side of the marked line and rest on the base of the vise. Tighten the jaws of the vise to secure the assembly in place. Do not apply too much pressure or you will distort the bearings, preventing the bolt from rotating. Secure the bolt in place with a nut followed by another to lock the first nut.

Step 2

Determine the desired diameter of the feed wheel and insert a drill bit with the same size upside down into the chuck. This drill bit acts as a gauge for positioning the vise onto the table of the drill press. Place the vise onto the table and position it so that the drill bit touches the side of the bolt and is aligned with the mark. Secure the vise in place by using a clamp and remove the drill bit from the chuck. Install the tap and apply a liberal amount of cutting oil to the cutting area.

Step 3

Turn on the drillpress and slowly lower the spindle until the tap begins to contact the bolt. Pressure is applied between the bolt and tap by slowly lowering the spindle and increases with the taper of the tap. This process requires a bit of finesse, and you must be careful not to lower the spindle too quickly. Apply enough pressure to engage the tap with the bolt so that the bolt begins to rotate as the teeth are cut. The key to this process is to work slow enough so as to not flex the tap but quick enough so that the blot turns as the tap cuts. Continue this process until the desired threads are formed. Remove the bolt from the holder and wipe off any filings and remaining cutting oil. The hobbed bolt is now ready for action. Install it into your extruder and start printing!

Learning in a Makerspace

8

Education is evolving and the idea of teaching with micromanufacturing technology is now a reality. Students are more responsive and retain more knowledge when they can connect the class work to the world in which we live. Yet students in the K–12 public education system are hammered with theoretical concepts that lack relevancy beyond the classroom, or are not taught in a way that makes them relevant. It is really not surprising that a good percentage of students lack the ability to apply their classwork to physical applications. It is imperative that we establish avenues for students to learn through hands-on exploration. This age of the Makerspace will serve as the catalyst for inspiring future minds in engineering and will act as an outlet for today's most creative individuals Figure 8-1.

Makerspace Education

Public school systems need Makerspaces. No longer should a school's poor financial resources limit access to the equipment necessary for higher-level exploration. It is now possible for a classroom—even one with a very small budget—to provide an environment conducive for utilizing and learning prototyping, robotics, electronics, microprocessors, and the like. Initially, the most logical location for a school-based Mak-

Figure 8-1. *An apple for the teacher?*

erspace is within the technology education and engineering classrooms. The staff have been trained on managing students in a workshop environment and the addition of Makerspace technology in the classroom is a natural fit.

This chapter's labs are designed to provide a starting point for getting students more involved in the technical aspects of your curriculum. Each K–12 lab activity harnesses the power of Makerspace technology in a way that allows the students to be more involved in the learning process. It helps to eliminate "black box" technology and gives the students more responsibility and ownership of their work. Each section contains labs categorized as a "Technical Lab," an "Activity," or a "Learning Tool." Technical Labs contain further investigation after the lab is complete. Activities do not contain an investigation and can serve as the starting point for a more technical lab. Learning Tools are tools that instructors can use in the classroom to help support theoretical concepts.

The labs are broken into three categories:

1. Kindergarten through 5th Grade: The teacher makes the parts using Makerspace technology and walks the students through each step of their use.

2. 6th through 8th Grade: The teacher helps the students make the parts, and the students then use them.

3. 9th through 12th Grade: The teacher demonstrates the equipment's use and the students design and make the parts.

Don't Limit Yourself!

Just because each lab focuses on different aspects of the science, technology, engineering, art, and math classroom doesn't mean they are not for you. Our schools spend too much time dividing students' attention in a way that tricks them into thinking these content areas don't, or sometimes shouldn't, overlap. In fact, programs benefit immensely when students are shown the connections between these areas and all of the areas in which they study.

Making and Makerspaces embrace collaboration and sharing of all kinds, especially when it comes to education. Read through each and find ways to inspire our youth to pursue not what is expected of them, but rather what excites them.

Achieving "STEM" Focus

Today's increased interest in science, technology, engineering, and math (STEM) courses was intended to boost interest in our technical work-force. The irony of this movement derives from the fact that the only required courses in STEM are science and math. And being required courses, they have required curriculum with standards testing to verify student comprehension. Updating the curriculum requires imaginative teachers who can both inject new and interesting topics into the coursework while hitting all of their benchmarks.

What About Art?

Art in STEM? Preposterous! Understanding the fundamental concepts of art in the design and creation of physical objects is imperative to success in STEM fields. Technology and engineering courses tackle design through both physical and software-based illustration. Courses such as basic technical drawing and architecture teach students to use manual drafting tools to visually represent their designs. Students can convey their ideas and reflect on possible design changes, thus making their designs more feasible. With computer-aided design tools such as AutoCAD, Solidworks, and AutoDesk Inventor, students can design the parts and assemblies that can be directly translated into real-world objects. Using tools such as these, students can make the connection between the virtual and physical worlds and see how design affects someone's ability to actually build something.

The Rhode Island School of Design has an excellent website dedicated to information and news concerning the integration of art and design into STEM (http://bit.ly/14ZWDbw) education.

So, all in all, it really ends up on the quality of the program and the emphasis the individual teacher puts on the integration of art into their curriculum. Art is incredibly important and more effort should be made in education to excite students about art, music, dance, theater, and so on. STEM education is designed to help increase excitement and interest in the fields of science, technology, engineering and math, all of which can benefit from the creativity art courses promote.

Makerspace Science

The science classroom offers a tremendous potential for utilizing Makerspace technology. The following labs are designed to enhance a student's experience within a science classroom by introducing common topics through hands-on investigation:

- "Lab #1: Making a Lemon Battery" on page 235
- "Lab #2: Making a Solar Cell" on page 237
- "Lab #3: Making a Microscope" on page 240

Lab #1: Making a Lemon Battery

Subject
Science

Grade Level
K–5

Lab Type
Technical Lab

Estimated Class Time
45 Minutes

Objective
This lab focuses on developing and applying an understanding of conventional battery technology. Through hands-on investigation and experimentation, students will explore different methods for producing low-power biological batteries and measuring their power output.

This project requires that safety glasses be worn throughout its entirety.

Overview
The earliest form of the battery consisted of a clay pot, a copper cathode and zinc anode, and a simple electrolyte solution. Although the exact uses of the battery are still a mystery, we can replicate this transformative technology in the classroom. A low-power battery can be safely made by modeling the use of copper and zinc and replacing the pot with a lemon (Figure 8-2).

Figure 8-2. *The lemon battery.*

When the zinc is suspended in an electrolyte solution—in this case, water and citric acid—small particles of zinc dissolve into the electrolyte. This process frees electrons that can power a very small electronic device such as an LED. Now, if the copper is also placed into the same electrolyte solution and the LED is connected between the two plates, a chemical reaction occurs that frees yet more electrons. You can measure the amount of energy produced by this simple battery by using a multimeter and calculate just how big of a device you can power. Who knows, maybe you could run your school on lemons! (Or, as the old saying goes: when life gives you lemons, power your school with them.)

The LED used in this lab requires 1.5 V and 20 mA of current to fully illuminate, and on average a good lemon battery will produce 0.5 VDC and 10 mA. Because the lemon battery does not produce enough power on its own to illuminate the light, more than one is needed. Multiple lemon batteries can be configure either in series or in parallel to meet the power requirement.

Connecting the batteries in series (Figure 8-3) increases the overall voltage. Each battery's voltage is added but the current stays the same. To configure lemon batteries in series, connect the positive terminal on one battery to the negative terminal on another. This is known as "daisy chaining," and you can continue adding more batteries in this fashion until the necessary voltage is achieved. The free terminal on the first and last batteries become the battery bank's positive and negative sides.

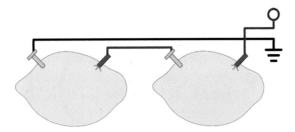

Figure 8-3. *Lemons connected in series will increase the voltage, but the current stays the same.*

Conversely, connecting the batteries in parallel (Figure 8-4) increases the overall current, but the voltage remains the same. To configure lemon batteries in parallel, connect the positive terminals and negative terminals of two or more batteries together. You can continue adding batteries like this until the necessary current is achieved.

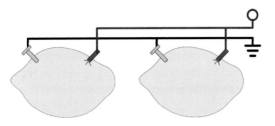

Figure 8-4. *Lemons in parallel increase the current, but this time the voltage stays the same.*

Materials

Materials List		
Item	**Quantity**	**Source**
Lemon	1	Grocery store

Materials List		
Copper penny	1	Pockets, under seat cushions, and so on
4D zinc-coated nail	1	Home improvement store
Alligator test leads	2	Electronics distributor
5 mm LED	1	Electronics distributor
Plastic knife	1	Grocery store
1 × 1 in 200-grit sandpaper	1	Home improvement store

Makerspace Tools and Equipment
Multimeter
Safety glasses

Setup
Step 1

Begin by placing your lemon on a stable work surface and gently roll it back and forth. This process will help produce the electrolyte base you require to make a good battery by releasing all of the internal citric acid-containing juices.

Step 2

Gently clean the surface of the penny with your sand paper so that it is bright and shiny. This process removes oxidation that will reduce the power of the battery. Now, using the plastic knife, carefully make a cut through the lemon's skin about 1/2 in wide and 1/2 in from the end of the lemon. Insert the penny so that approximately 1/8 in. is left exposed.

Step 3

Insert the zinc-coated nail into the lemon, about 1/2 in from the opposite end of the lemon as the penny.

Step 4

With the help of your instructor, configure the multimeter to measure volts DC (VDC). Touch one lead to the penny and one to the nail. If you see numbers on the screen, congratulations—you just made a battery out of a lemon! Now, lets take some measurements (Figure 8-5).

Figure 8-5. Configuring multiple lemon batteries in series and in parallel changes the amount of available voltage and current.

Investigation

The electric capability of your battery is measured in terms of voltage and current. To understand this capability, you will need to collect data that reflects its power output potential by using a multimeter.

Determine Polarity and Voltage

1. A battery has two sides, positive and negative. Following the same procedure as described in step 4, touch one of the multimeter's leads to the penny and one to the nail. If the numbers on the screen are negative, your leads are connected backward. Flip your lemon around and try again.

    a. Is the penny the positive or negative side?

    b. Is the nail the positive or negative side?

    c. How much voltage is produced?

Determine the Current

1. With the help of your instructor, configure the multimeter to measure milliamperes (mA). Touch the multimeter's red lead to the positive end of the battery and the black lead to the negative.

a. How much current is available?

Lighting an LED

1. Connect one alligator test lead to the end of the nail and one to the exposed part of the penny. You'll use these leads to connect your LED to the lemon battery. Now, connect the negative side of your battery to the negative side of the LED as indicated with a black stripe and the positive side of the battery to the positive side of the LED.

    a. Did the LED light up?

2. Remember, the LED requires approximately 1.5 V and 20 mA to glow. If your battery produces 0.5 V and 10 mA, how many batteries in series and parallel would you need in order to light the LED?

    a. Number of batteries in series

    b. Number of batteries in parallel

3. Working with one or more other students, configure your batteries to produce enough energy to illuminate the LED.

    a. Number of batteries in series

    b. Number of batteries in parallel

    c. Were you able to illuminate the LED?

Lab #2: Making a Solar Cell

Subject
    Science

Grade Level
    6–8

Lab Type
    Technical Lab

Estimated Class Time
    1+ Hours

Objective

This lab demonstrates the production and use of a light-sensitive semiconductor in the formation of a primitive solar cell. Copper is used to produce a naturally occurring semiconductor, cuprous oxide, which forms on the surface of copper when exposed to high heat.

This project requires that safety glasses be worn throughout its entirety.

Overview

Solar cells, or photovoltaics, are made out of photosensitive semiconductors that convert light energy into electricity in a process known as the *photoelectric effect*. This process describes the interaction between light energy and the materials that make up the individual cell. Although commercial solar cells are made by using harmful chemicals and require sophisticated equipment for assembly, you can make a simple low-cost solar cell (Figure 8-6) in your lab with only a few materials.

Figure 8-6. *Measuring the output of the solar cell.*

Materials

Materials List		
Item	**Quantity**	**Source**
Table salt	1 tbsp	Local grocery

Materials List		
White vinegar	1 cup	Local grocery
Distilled water	1 cup	Local grocery
Small plastic cup	1	Local grocery
Thin copper sheet	2	Home improvement store
Alligator test leads	2	Electronics distributor

Makerspace Tools and Equipment
High-temperature hot plate
Micrometer
Nonflammable work surface
Multimeter
Safety glasses
Tweezers

Setup

Step 1

Place the hot plate on a nonflammable surface and preheat it to 500°F. Ensure that there is enough free area around the hot plate for placing the heated copper.

Step 2

Cut the copper sheet into 1 × 1 in squares. Each solar cell will require two pieces. Remove the existing oxidation and dirt from the surface of the copper using a cleaning solution made with vinegar and water. In a cup, mix 1 cup of vinegar and 1 tbsp of table salt until the salt has completely dissolved.

Step 3

Create a grid on a piece of printer paper and assign labels for the rows and cells so that there are enough spaces for the number of students in the class.

Procedure

Step 1

Measure the surface area of the copper pieces using the dial caliper. This measurement will be used to determine the solar conversion efficiency.

Step 2

Using the supplied cleaning solution, carefully remove the insulating oxide and dirt on the surfaces of both of your copper pieces. Add a small amount of the solution to the paper towel and rub the copper until shiny. One piece is going to act as the positive and the other the negative.

Step 3

Locate one grid space on the provided sheet and place only one of your pieces in the cell. The instructor should place the piece onto the hot plate. As the copper heats up (Figure 8-7), observe the formation of the red cuprous oxide and then the resulting formation of black cupric oxide, an insulator that will be removed at the end of the heating cycle. When the cupric oxide has completely covered the surface of your piece, the instructor turns off the hot plate and allows the copper to cool down. During this time the cupric oxide should begin to flake off. Let the copper cool before handling.

Figure 8-7. *As the copper is heated, a rainbow of cuprous oxide forms, followed by a layer of dark cupric oxide.*

Step 4

Carefully clean away the resulting cupric oxide with sandpaper to expose your cuprous oxide semiconductor. Do not remove the multicolored cuprous oxide (Figure 8-8) during this cleaning process!

Figure 8-8. *You should have two pieces of copper to complete the cell, one with cuprous oxide and one without.*

Step 5

Prepare a solution of distilled water and 1/4 tsp of salt in your plastic cup, making sure that all of the salt has dissolved. Carefully attach your test leads to each piece and submerge them into the salt solution. Take care that the copper pieces do not touch.

Investigation

Now that you have successfully made a primitive solar cell, it's time to determine how much energy it can generate. Using a multimeter, you will be conducting a series of tests to determine the polarity, voltage output, current capability, and overall efficiency.

Determine Polarity and Voltage

1. A solar cell has two sides: one positive and the other negative. Connect your multimeter's negative (black) lead to one test lead and the positive (red) lead to the other test lead. Configure your multimeter to measure VDC and adjust the scale accordingly. If the numbers on the screen are negative, your leads are connected backward.

   a. Is the penny with cuprous oxide the positive or negative side?

   b. How much voltage is produced?

Determine the Current

1. Configure the multimeter to measure current and adjust the scale accordingly.

   a. How much current is available?

Determine the Efficiency

The sun produces approximately 1,000 W/m^2 of light energy. Determine your solar cell's energy conversion efficiency by dividing your W/m^2 by that of the sun's. Remember, Watts = Current × Voltage.

   a. How efficient is your solar cell?

Lab #3: Making a Microscope

Subject
Science

Grade Level
9–12

Lab Type
Learning Tool

Estimated Class Time
1.5 Hours

Objective
The microscope is one of the most sensitive tools found in the science classroom. Its glass slides and clean lenses require careful handling to give students a glimpse into the microworld around us. With a little help from a Makerspace, the students can create a simple microscope using only a few cents' worth of materials.

This project requires that safety glasses be worn throughout its entirety.

This project works with very hot materials and should only be conducted under the supervision of a qualified individual.

Overview

During the late 1600s, Anton van Leeuwenhoek began developing single-lens microscopes (Figure 8-9) that were capable of magnifying objects to more than 200 times their actual size. This incredible invention led to the discovery of numerous microorganisms and cell types.

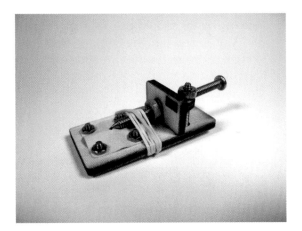

Figure 8-9. *The completed microscope.*

Materials

You can find the files for this project at Thingiverse.com (*http://bit.ly/1asKF21*).

Materials List			
Item	**Quantity**	**Cost**	**Source**
1 × 2 × 1/8 in plywood	1		Craft store
1/8 × 2 in wooden dowel	1		Craft store
4-40 × 2 in pan-head machine screw	1		Home improvement store
4-40 Nut	2		Home improvement store
#4 Washer	1		Home improvement store
5 mm × 6 in glass stirrer	1		Scientific supply company
Rubber band	1		Home improvement store

Makerspace Tools and Equipment
Countersink bit
Heat-resistant gloves
Laser engraver
Wire strips
Wood glue
Propane torch or Bunsen burner
Safety glasses

Setup
Step 1

Open the *scope.dxf* file in your CAD software suite and create an array for the quantity of microscopes that you want to make. Follow your laser cutter's operating procedure for 1/8 in plywood. Position the sheet of plywood on the cutting bed, send your drawing to the machine, power on the exhaust fan and air assist, and begin cutting.

Procedure
Step 1

Put on your safety glasses and heat-resistive gloves. Ignite the torch and adjust the flame so that it is about 2 in length. Grip the stirrer approximately 1 in from each end and insert the middle of the rod into the flame. Ensure that the rod is positioned at the top of the blue cone, which is the region with the highest temperatures.

Step 2

Gently roll the rod back and forth with your fingers to evenly distribute the heat. Continue this process until the glass begins to deform and wilt (Figure 8-10). When the rod is noticeably deformed, remove it from the heat and immediately pull the ends so that a 1 mm filament is formed. Pass the middle of the filament back into the flame and pull again to make two 2 mm filaments.

Figure 8-10. *Roll the rod in the flame until it begins to wilt and then carefully pull it into a 1 mm filament.*

Step 3

Grip the thick end of one of the stirrers and carefully insert the 2 mm end into the flame at a 45-degree angle. The objective here is to form a glass ball at the end of the filament. Gradually feed the filament into the flame until the ball grows to approximately 3 to 4 mm in diameter (Figure 8-11).

When you have achieved this size, remove the filament from the flame, turn off the torch, and hold the rod perpendicular to your work surface until the lens cools (Figure 8-12). Clip the lens off of the filament by using the wire strippers so that 1/2 in of filament is left attached to the lens.

Figure 8-11. *Remove the glass from the flame as soon as you create a 3 to 4 mm ball lens. If the lens grows too big, it might fall off of the filament.*

Figure 8-12. *Hold the rod vertically as the lens cools to help align the filament and better shape the lens.*

Do not quench the lens in water because the thermal shock will shatter the glass.

Step 4

Use the countersink bit to recess the lens holder on both sides of the microscope body. Be careful that the 1 mm hole does not get widened. Position the lens into the lens holder and secure it in place with a drop of adhesive.

Step 5

Glue the mount onto the focus support. After the glue dries, insert the support into the microscope body and secure it in place with the focus screw. Sand the end of the dowel to produce a section approximately 30 degrees relative to the shaft. This will act as the specimen holder.

Step 6

Insert the dowel into the focus support and secure it in place with a rubber band. This band also helps prevent the dowel from rotating. Attach your specimen to the holder and hold the opposite end of the microscope to your eye. Look through the lens and focus the holder accordingly. This process works best if you backlight the specimen with a bright light source (Figure 8-13).

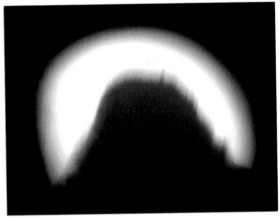

Figure 8-13. *Using the microscope takes a bit of finesse. Test different materials to see which works best.*

Makerspace Technology

Technology Education was designed to be a Makerspace. Most technical education classrooms are designed to function as workshops and can support projects starting with CAD design all the way to a completed prototype. By introducing these into the classroom, students can experience the full potential Makerspace technology has to offer:

Lab #4: Making a Water Rocket and Launcher

Subject
> Technology

Grade Level
> K–5

Lab Type
> Technical Lab

Estimated Class Time
> 1.5 Hours

Objective
> Everybody loves rockets, especially elementary school students who marvel at the simple construction and launch of a water-propelled spacecraft model. Exposing students at this age to the fundamentals of physics will help them stay excited and potentially inspire them to become our future engineers and space explorers.

This project requires that safety glasses be worn throughout its entirety.

Overview

Newton taught us that for every action there is an equal and opposite reaction. Rockets use this very principle. Stored inside a commercial rocket is a large amount of fuel. This fuel is ignited, and the rapidly expanding gases exit the rocket with a tremendous amount of force, propelling the rocket in the opposite direction. Rather than have elementary school students handle volatile rocket fuels and dangerous propulsion systems, you can re-create the basic concepts behind rocket technology safely in the classroom (Figure 8-14).

Figure 8-14. *Water-propelled rocket ready for launch.*

Materials

You can find the files for this project at Thingiverse.com (*http://bit.ly/1d2Z0UU*).

Materials List		
Item	**Quantity**	**Source**
Empty 2-liter bottle	1	Grocery store
Pipe tape	1	Home improvement store
24 in cable tie	2	Home improvement store
12 × 24 in corrugated plastic sheet	1	Home improvement store
2 × 24 in PVC pipe	1	Home improvement store
2 in PVC pipe cap	1	Home improvement store
Modeling clay	1	Craft store
1/4-20 × 1/2 in eye bolt	1	Home improvement store
1/4 in washer	1	Home improvement store
1/4-20 nut	1	Home improvement store
18 in string	1	Home improvement store

Materials List

1/4 in NPT quick-release female coupler with female thread	1	Home improvement store
1/4 in NPT male plug with male thread	1	Home improvement store

Makerspace Tools and Equipment

3D Printer

1.5 in hole saw

1/4 in drill bit

Air compressor w/ hose

Band saw

Hand drill

Packing tape

Ruler

Safety glasses

Scissors

Making the Launcher
Step 1

Start by drilling a 0.5 in diameter hole in the center of the PVC endcap. Wrap a small amount of thread-sealing tape around the threads of the male NPT plug. Pass the threaded portion of the plug through the hole and secure it with the quick release coupler. Attach the air hose to the male plug, and the top of the launcher is set.

Step 2

Secure the PVC pipe in a vise and drill a 1.5 in diameter hole in the side of the pipe 6 in from the end by using a hole saw. Drill a 1/4 in hole 1 in above the 1.5 in hole. This hole will be used to secure the eye bolt (Figure 8-15). Taper the other end of the pipe by using the band saw. This taper will allow for the pipe to be driven into dirt, securing it in place during a launch.

Figure 8-15. *Drilling a hole in the side of the PVC pipe will allow for the air hose to pass through.*

Step 3

Secure the eye bolt into the 1/4 in hole using a washer followed by a nut. This bolt is used to guide the launch cord. Pass the air hose through the 1.5 in hole and press on the endcap. Tie one end of the string around the body of the NPT coupler and pass it through the eye hole. Tug on the string to ensure that the coupler releases, and the launcher is complete (Figure 8-16)!

Figure 8-16. *NPT coupler-based water rocket launcher.*

Making the Nozzle, Fins, and Nosecone
Step 1

Power up your 3D printer and start the control software. Import the *BottleRocketNozzle.stl* file and position onto the virtual build surface. The tapered end of the nozzle must face away from the build surface because the design contains overhangs. Begin by printing only one nozzle to ensure its quality and then print an array of the necessary quantity. This will save you the headache of finding out after an 8-hour print that your nozzles don't fit.

Step 2

Configure your build properties for 100 percent infill and send the layout to the printer. When the nozzle has finished printing, perform a fit check (Figure 8-17) on an unpressurized male quick connect coupler. The nozzle should be snug when seated and pop off easily when the release mechanism is activated.

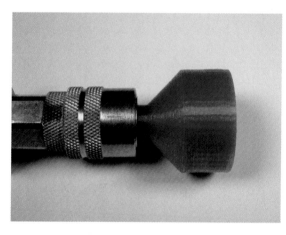

Figure 8-17. *Test fit the nozzle after printing. It should be snug enough to make a seal but loose enough to pop out of the connector when released.*

Step 3

Pressure test your nozzle by inserting an O-ring into the nozzle and firmly fastening it to an empty bottle. Adjust your air supply to 40 PSI and connect the bottle. You should now hear minimal air leakage. If leakage is excessive, try scaling down the STL by 1 or 2 percent and try again until the problem is resolved.

Step 4

Open the *bottleFin.dxf* file in your 2D CAD software and create an array for the quantity of fins that you want to make. Typically, the rockets will use 3, but 4 or more might be desired for experimentation. Follow your laser cutter/engraver's operating procedure for configuring the machine for cutting. Position the sheet of plywood on the cutting bed, send your drawing to the machine, power on the exhaust fan and air assist, and begin cutting.

Procedure
Step 1

Make sure all labels have been removed from the outside of the bottle and it is free of major scratches or cracks. Use a doorjam to draw a straight line down the side of the bottle with a marker. Wrap a string or piece of paper around the bottle and mark at the point of overlap. Fold the string into thirds and mark the points with a marker. Reattach the string to the bottle and transfer the two new marks. Again, use the doorjam to draw two new lines down the side of the bottle. You should now have three evenly spaced lines drawn on your bottle that will act as guides for aligning your fins.

Step 2

Line up and attach one fin at a time to the previously marked positions using tape. Verify that the fins are positioned perpendicular to the bottle and that the tape is pressed firmly onto the surface of the fins and bottle. Repeat this step for the remaining fins.

Hold the fins in place using the large cable ties (Figure 8-18). If you want, apply a small amount of hot glue along the interface between the fins and the bottle. Use a low temperature hot-glue gun because the high temperature guns will melt the bottle.

Step 3

Roll up the supplied clay into a ball, press it onto the base of the 2-liter bottle, and then secure it in place with a piece of tape. This clay will help to stabilize the flight of your rocket by moving its center of mass closer to the nose.

Figure 8-18. *Securing the fins.*

Step 4

Now that you have created your water rocket, it is time to decorate it. Use stickers, paint, and markers to personalize your rocket so it stands out from the rest. Finally, it's time to launch.

Step 5

To launch your rocket, drive the launcher into the ground so that more than half of the 24 in PVC tube is underground. Remove the nozzle from the bottle and fill it 1/4 full with water. Reattach the nozzle and connect it to the launcher. Confirm that everyone is wearing safety glasses and is standing a safe distance away from the rocket and then, using the air compressor, pump up the rocket to no more then 40 PSI. Start your countdown: 4… 3… 2… 1… LAUNCH! Pull the string and watch the rocket fly!

Lab #5: Making a 3D Printed Robot

Subject
Technology

Grade Level
6–8

Lab Type
Learning Tool

Estimated Class Time
3+ Hours

Objective
Working with robotics is a marvelous way to expose students to multiple technology and engineering concepts operating in sequence. Robots require both software and hardware to function, and when they work together, various levels of "intelligence" can be simulated.

This project requires that safety glasses be worn throughout its entirety.

Overview

This lab bridges the gap between the virtual world of software and physical hardware. The students are required to implement the engineering design process to the design and construction of a simple two-wheeled robot (Figure 8-19) that is capable of autonomously completing tasks.

Figure 8-19. *The completed robot.*

Materials

You can find the files for this project at Thingiverse.com (*http://bit.ly/13GMczs*).

Materials List		
Item	**Quantity**	**Source**
4-40 × 1/4 in machine screws	4	Home improvement store
4-40 × 1/2 in machine screws	2	Home improvement store
4-40 nut	6	Home improvement store
#4 washer	6	Home improvement store
4-40 standoff	4	Electronic supply store
Tamiya 70168 gearbox kit	1	Hobby store
Glass marble	1	Craft store
L293D	1	Electronic supply store
Pre-stripped breadboard jumper wire	1	Electronic supply store
Male-to-male jumper wires	1	Electronic supply store
2 ft Red and black 22-gauge wire	1	Electronic supply store
Arduino	1	Electronic supply store
Small breadboard	1	Electronic supply store
9-Volt Battery	1	Electronic supply store
9-Volt Battery snap connector	1	Electronic supply store

Makerspace Tools and Equipment
3D Printer
Band saw
CAD software
Solder
Soldering iron

Setup
Step 1

Open the *robotPlate.stl*, *wheelRight.stl*, and *wheelLeft.stl* files in your 3D printer's control software and position it onto the virtual build surface. Process the build and send it to the 3D printer for fabrication. This file contains both the chassis and the wheels for the robot. While the chassis is being printed, assemble kits of electronic components for distribution to the students.

Step 2

Open the gearbox kit and solder the four motor wires to the motor's terminals. Be careful not to apply too much heat because the motor is easily damaged. Disconnect the soldering iron when you're done.

Procedure
Step 1

Assemble the gearbox per the manufacturer's instructions. Position the four Arduino mounting nuts into the sockets on the chassis and attach the 4-40 standoffs. Remove the adhesive from the back of the breadboard and position it on the front of the robot chassis.

Step 2

Attach the gearbox to the chassis by using the two 1/2 in 4-40 bolts (Figure 8-20). Secure it in place with a washer and a nut. Attach the wheels to the gearbox and press the marble into the caster. Using the four 4-40 machine screws, attach the Arduino to the 4-40 standoffs. To avoid damaging the board, do not overtighten these screws.

Figure 8-20. *Attach the gearbox.*

Step 3

Populate the breadboard by first seating the IC on the board and then securing the wires in place.

Use the male-to-male jumper wires to make connections back to the Arduino, following the color codes to ensure proper wiring. Use Figure 8-21 and Table 8-1 to connect the L293D to your Arduino.

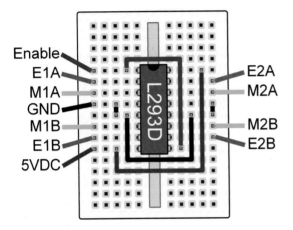

Figure 8-21. *L293D breadboard layout.*

Table 8-1. *Motor Controller/Arduino/Motor Connection*

Motor controller	Arduino	Motor
Enable	D3	
E1A	D4	

Motor controller	Arduino	Motor
M1A		Motor 1 red
GND	GND	
M1B		Motor 1 black
E1B	D5	
5 VDC	5 VDC	
E2A	D6	
M2A		Motor 2 black
M2B		Motor 2 red
E2B	D7	

Step 4

Connect the robot to your computer and open the Arduino IDE. Copy Example 8-1 into your sketch and upload it to your robot.

Step 5

Snap the 9-volt battery into the holder and connect the snap connector. Your Arduino should now be powered and will run the robot test code. The robot should move forward and then backward when the test code is executed. If it does not, disconnect the battery and swap positions of M2A/M2B. This will reverse the direction of Motor 2.

Example 8-1. **Basic robot**

```
const int enable = 3; //defines enable pin number
const int e1a = 4; //defines e1a pin number
const int e2a = 5; //defines e2a pin number
const int e1b = 6; //defines e1b pin number
const int e2b = 7; //defines e2b pin number

void setup() {
  pinMode(enable, OUTPUT); //sets pin states to output
  pinMode(e1a, OUTPUT);
  pinMode(e1a, OUTPUT);
  pinMode(e2b, OUTPUT);
  pinMode(e2a, OUTPUT);

  digitalWrite(enable, HIGH); //enables the H-Bridge

}

void loop() {
  digitalWrite(e1a, HIGH); //turns both motors in one direction
  digitalWrite(e1b, LOW);
  digitalWrite(e2a, HIGH);
```

```
  digitalWrite(e2b, LOW);
  delay(2000);
  digitalWrite(e1a, LOW); //turns both motors in the opposite direction
  digitalWrite(e1b, HIGH);
  digitalWrite(e2a, LOW);
  digitalWrite(e2b, HIGH);
  delay(2000);
}
```

Lab #6: Making a Brushless DC Motor

Subject
Technology

Grade Level
6–8

Lab Type
Learning Tool

Estimated Class Time
1.5 Hours

Objective
The electric motor has been around since the early 1800s and continues to power our world today. This lab focuses on developing an understanding of basic electromagnetism and the technology that governs the operation of brushless DC motors. This apparatus can be used as the basis for developing curriculum that focuses on the creation, application, and analysis of brushless DC motors.

This project requires that safety glasses be worn throughout its entirety.

Overview
Regardless of the type, DC electric motors contain two fundamental components: the rotor and the stator. The rotor is responsible for converting electric energy into rotational motion, and the stator acts to provide a mechanical coupling to the rotating armature. Although both brushed and brushless DC motors convert electrical energy into rotational energy, they contain only a few common components. A brushed DC motor is made up of the following:

Armature Core
The armature core is made up of ferrous plates that are bonded together to make a solid stack. This stack usually consists of three poles on which the armature winding is wound.

Armature Winding
The armature winding consists of insulated wire wound around the poles of a magnetic armature core. When energized, each pole becomes an electromagnet pulling the armature in line with the field magnet.

Axle
Supports the armature and commutator with either bushings or bearings, allowing them to spin freely.

Brushes
Consist of either carbon blocks or metal fingers that connect the DC power source to the commutator.

Commutator
Consists of metal plates equal to the number of electromagnets attached to the armature. These metal plates are connected in series with the each end of the coils and act as the electric bridge between the armature and the brushes.

Field Magnet
Consists of two or more permanent magnets that work in conjunction with the polarity of the armature to produce the force behind the rotational motor's motion.

However, a brushless DC motor (Figure 8-22) only contains the following:

Axle

Supports the rotating rotor and is coupled to the stationary coil assembly with a series of bearings.

Control Electronics

Also known as an Electronic Speed Control, or ESC, the control electronics are responsible for switching on and off the driving coils, pulling the rotor in line with the electromagnetic field.

Driving Coils

Brushless DC motors typically have three or more driving coils. When energized, the coils pull the rotor in a series of controlled pulses.

Permanent Magnet Rotor

These are typically made of rare earth magnets and can be positioned internal or external to the motor. This "out-runner" and "in-runner" design makes sophisticated cooling methods possible and allows the motor to perform with greater than 90 percent efficiency.

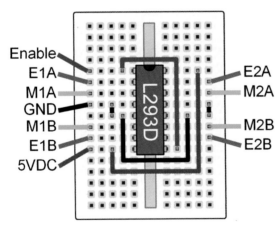

Figure 8-22. *The completed motor.*

Materials

You can find the files for this project at Thingiverse.com (*http://bit.ly/1bU9SAs*).

Materials List

Item	Quantity	Source
1/8 in plywood Sheet	1	Home improvement store
1/4-20 × 1 in bolt	2	Home improvement store
1/4-20 nut	2	Home improvement store
#4 washer	2	Home improvement store
Breadboard	1	Electronic supply store
10 ft 30-gauge coil wire	2	Electronic supply store
470 Ohm resistor	2	Electronic supply store
47k Ohm resistor	2	Electronic supply store
220 uF capacitor	2	Electronic supply store
TIP120 transistor	2	Electronic supply store
2N2222 transistor	2	Electronic supply store
Pre-stripped breadboard jumper wire	1	Electronic supply store
6 in 22-gauge solid core wire	4	Electronic supply store
9 V battery	1	Electronic supply store
9 V battery snap connector	1	Electronic supply store
1 in dia. doughnut magnets	2	Craft store
1/8 in dia. drinking straw	1	Grocery store

Makerspace Tools and Equipment

Laser engraver	
Wire strips	

Setup
Step 1

Open the *brushlessDC.dxf* file in your CAD software suite and create an array for the quantity of motors that you want to make. Follow your laser cutter's operating procedure for configuring the machine for cutting. Position the sheet of plywood on the cutting bed, send your drawing to the machine, power on the exhaust fan and air assist, and begin cutting.

Step 2

Prepare kits containing the remaining components into sandwich bags for easy distribution to the students.

Procedure

Step 1

Assemble the controller electronics following the provided diagram and the illustrations on the back of the circuit board (Figure 8-23). This circuit is a simple *astable multivibrator*.

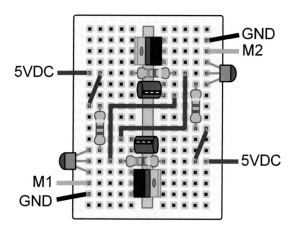

Figure 8-23. *Breadboard layout for brushless DC motor controller.*

Step 2

Use the wire snaps to cut two 10 ft lengths of coil wire. Pass one bolt through the motor's vertical support and add a washer and nut to the threaded side. Leave 1/4 in of thread exposed under the screw's head and tightly wrap all 10 ft of coil wire around the threads. Try to keep the coil wire as tight as possible. When complete, secure it in place with a drop of hot glue.

Classroom floor tiles are often 1 × 1 ft and make for a quick measurement tool.

Step 3

Assemble the rotor by sliding the magnet mounts onto the center of the axle and glue it in place with a drop of super glue. Add the magnets so that the poles align and secure them in place with a drop of super glue. Place an 1/8 in piece of drinking straw into the axle

mounts. This acts as a bearing surface, allowing the axle to spin more freely (Figure 8-24).

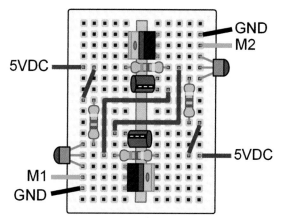

Figure 8-24. *The assembled rotor. Make sure that the coils do not interfere with the rotor.*

Step 4

Test fit the motor structure, and if you're satisfied, glue together the pieces by using wood glue; allow to dry completely. Check that the coils are facing toward the inside of the structure and that the rotor spins freely. Slide the two wooden washers over the ends of the axle and glue them in place with super glue. These prevent the rotor from "wandering" while in motion. Verify that the rotor spins freely after gluing.

Step 5

Attach the coil wires to the control board by carefully sanding away the insulation at the ends of the wire, securing them into the breadboard. If the coil wires do not fit nicely into the breadboard, they can be soldered to short 22-gauge jumper wires that better fit in a breadboard. Energize the control board by attaching the 9-volt battery, and the rotor should begin to shake. Give the axle a quick spin with your fingers to start the motor. If it doesn't work, try switching the polarity on one coil and ensure that your circuitry was assembled correctly.

Makerspace Engineering

Engineering bridges the gap between the concepts covered in Math and the structures constructed in Technology Education. These principles can be demonstrated and validated by using Makerspace technology. The following labs are designed to develop problem-solving skills through the design, construction, and analysis of engineering concepts:

- "Lab #7: Making a Sail Car"
- "Lab #8: Making a Wind Turbine"
- "Lab #9: Making a Hot-Air Balloon"

Lab #7: Making a Sail Car

Subject
    Engineering

Grade Level
    K–5

Lab Type
    Activity

Estimated Class Time
    45 Minutes

Objective
    Giving elementary school students problem-solving challenges exposes them directly to the bridge between imagination and creation.

This project requires that safety glasses be worn throughout its entirety.

Overview

Problem solving is the most important skill a student can have. They can use it to conceptualize the solution to a task prior to any research, bookwork, or prototyping. It also gives them insight as to whether their research, bookwork, or prototype is headed in the right direction. This lab is designed to expose elementary school students to the trials faced when designing and constructing a wind-powered car (Figure 8-25). After the kit has been assembled, the students should be given the opportunity to test and redesign their solutions for best results.

Figure 8-25. *The completed car.*

Materials

You can find the files for this project at Thingiverse.com (*http://bit.ly/19K6lXT*).

Materials List		
Item	**Quantity**	**Source**
1/8 × 3.5 in wooden dowel	1	Craft store
1/8 × 3 in drinking straw	1	Grocery store
1/4 × 5 in wooden dowel	1	Craft store
3 × 5 in notecard	1	Craft store
Notebook paper	1	Craft store
Tape	1	Craft store
Cardboard	1	Craft store

Makerspace Tools and Equipment
Box fan
Hand saw
Hole punch
Laser engraver
Safety glasses

Setup

Step 1

Open the *windCar.dxf* file in your CAD software suite and create an array for the quantity of cars that you want to make. Remember to include one body and four wheels for each car. Follow your laser cutter's operating procedure for configuring the machine for cutting. Position the sheet of cardboard on the cutting bed, send your drawing to the machine, power on the exhaust fan and air assist, and begin cutting.

Step 2

While the cars are being cut on the laser, prepare the axles and straws by cutting them to length using your wire strippers, and cut the 1/4 in dowel using a saw. Gently sand the ends of the dowels to prevent splinters. Punch two holes in the sheet of paper 3 in apart by using a hole punch. These holes will be used to hold the sail in place. When the laser cutting has finished, assemble kits of parts in sandwich bags for distribution to the students.

Procedure

Step 1

Begin by attaching the two car sides onto the body by using small pieces of tape or glue. Pass the two straw pieces through the mounts on either side of the car and then secure in place with a piece of tape. Be careful not to bend the straws; this will prevent the wheels from turning.

Step 2

Assemble the mast by passing the 1/4 in dowel through the holes in the card and into one of the positioning holes. These holes will be used to determine the best position of the mast through testing. Secure the mast in place with a piece of tape.

Step 3

Place the wind car in front of the fan and turn it on. The car should now travel across the floor using the power of the breeze created by the fan. The performance of your wind car can be improved by changing the design of the sail. Use extra sheets of paper, a hole punch, and scissors to design new types of sails. Try to make the fastest possible car. Which design worked best?

Lab #8: Making a Wind Turbine

Subject
Engineering

Grade Level
6–8

Lab Type
Technical Lab

Estimated Class Time
4+ Hours

Objective
Generating electrical energy from renewable sources requires a lot of engineering and imagination. One of the simplest forms of energy that we can harvest is that of the wind. Whether installed off-shore, on mountain tops, or on open plains, wind turbines provide a low impact and effective method for converting wind energy into electrical.

This project requires that safety glasses be worn throughout its entirety.

Overview

Wind turbines are elegant machines that convert power from the wind into electrical energy. This is accomplished when the wind passes over the surface of the turbine's blades generating lift and rotational force. The rotating blades are connected to a generator that creates electrical energy when spun.

This lab utilizes inexpensive materials to create a small, low-power, horizontal axis wind turbine (Figure 8-26). You can measure the efficiency of the turbine by using a multimeter and can alter it by changing the blade's angle of attack as well as refining blade design.

Figure 8-26. *The completed turbine.*

Materials

You can find the files for this project at Thingiverse.com (*http://bit.ly/1d2ZfPK*).

Materials List		
Item	**Quantity**	**Source**
1/8 × 1/4 in × 36 in balsa strips	4	Craft store
1/8 × 3 in × 12 in balsa sheet	1	Craft store
1/8 × 3 in booden dowel	3	Home improvement store
DC electric motor with 2 mm shaft	1	Hobby store

Makerspace Tools and Equipment
Laser engraver
3D printer
Multimeter
Box fan
Vise
Hot-glue gun with glue

Making the Blade Hub
Step 1

Open the *WindTurbineHub.stl* file in your 3D printer's control software and position it onto the virtual build surface. Process the build and send it to the 3D printer for fabrication. Using a low build quality is acceptable here because it will decrease your build time. Assemble kits of parts for distribution to the students while the hubs are printing.

Making the Tower
Step 1

Print out a full scale copy of the *turbine.dxf* tower template and tape it to your work surface. Tape an overhead projector transparency over the template to prevent the glue from sticking.

Step 2

Use a razor to carefully align and cut out all of the beams and cross members as shown on the template. Tape down the beams between the joints, apply a small amount of glue to the ends of the cross members, and then glue them into place. After the glue has dried, repeat this process two more times. Glue the three tower sections together to form your wind turbine's tower (Figure 8-27). After the tower is dry, glue it to the hardboard base and attach the motor mount.

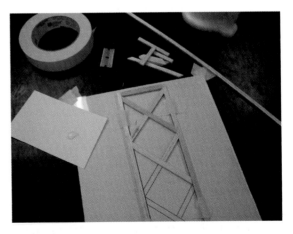

Figure 8-27. *Glue the pieces together on top of a piece of plastic sheet. This will prevent them from sticking to the drawing.*

Step 3

Using a small vise, press the hub onto the motor (Figure 8-28). Do not try to bore out the hub; this will prevent it from coupling to the motor's drive shaft. Secure the motor to the top of the turbine by using a small amount of hot glue.

Figure 8-28. *Use a vise rather than a hammer to attach the hub to the motor because the blows of a hammer will damage the motor's internal components.*

Making the Blades
Step 1

Open the *turbine.dxf* file in your CAD software suite and create an array of the number of blades that you want to make. Follow your laser cutter's operating procedure for configuring the machine for cutting. Position the 1/8 in balsa sheet on the bed. Send your drawing to the machine, power on the exhaust fan and air assist, and begin cutting.

Step 2

Attach each blade to the end of the dowels and secure them in place with two cable ties. When you have the blade position set, glue the blade to the dowel by using a small amount of wood glue. Connect the blades to their respective holes on the hub. If the blade twists on the wooden dowel, secure it in place with a small amount of hot glue (Figure 8-29). This hub is designed for 3 or 4 blades and will distribute the blades evenly.

Give your turbine a gentle spin and see if it works.

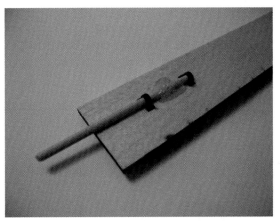

Figure 8-29. *Securing the blade with hot glue.*

Investigation
Wind turbines operate by converting the flow of air to a rotational force that drives a generator. This is accomplished through the research and design of blades that harness as much wind energy as possible while producing the least amount of resistance.

Measuring Output
1. Connect your multimeter's negative (black) lead to one test lead and the positive (red) lead to the other test lead. Configure your multimeter to measure VDC and adjust the scale accordingly. If the numbers on the screen are negative, your leads are connected backward.

   a. What happens to the voltage as the turbine is moved closer to the fan?

   b. What happens to the voltage as it is moved farther away?

Adjusting Angle of Attack
1. The angle of attack is the angle at which the blades are positioned relative to the path of the wind.

   a. How does a higher angle of attack affect rotation?

b. How does a lower angle of attack affect rotation?

Modifying the Blades

1. Modify the design of the blades to improve the speed of rotation. Produce different shapes, both wide and narrow.

   a. How do wide blades affect rotation?

   b. How do narrow blades affect rotation?

Lab #9: Making a Hot-Air Balloon

Subject
  Engineering

Grade Level
  9–12

Lab Type
  Technical Lab

Estimated Class Time
  3+ Hours

Objective
  This lab is designed to expose students to the effect of temperature on the density of a gas by constructing and flying a tissue-paper hot-air balloon. Through investigation, the students will make the correlation between volume, temperature, and lifting force.

This project requires that safety glasses be worn throughout its entirety.

Overview

Hot-air balloon technology has been around since the late 1700s and still amazes onlookers today. The fundamental design of a hot-air balloon is based around the Ideal Gas Law. This law states that the pressure and volume of a gas directly relates to its temperature and moles of gas:

$$PV = nRT$$

Where P equals the gas's pressure in Pa, V is the volume in m^3, n is the number of moles of gas, R is the ideal gas constant of 287.058 J/(kg × K), and T is the temperature of the gas in K. You can write this equation to illustrate the effect of temperature on the density of a gas:

$$\rho = P / (R \times T)$$

Where ρ represents the gas's density in kg/m^3, P is the pressure in Pa, R is the ideal gas constant of 287.058 J/(kg × K) and T is temperature in K.

You can use this equation to express the quantity of mass a hot-air balloon can lift based on the volume of the balloon and its temperature. This is known as the *balloon's buoyant force*. This force can be calculated by accounting for the difference in density of the air inside and outside of the balloon and its weight. So, will it float?

$$F_t = (\rho_a - \rho_b) \times V \times g$$

Where F_t represents the quantity of force generated by your balloon, ρ_a is the density of the air outside of the balloon in kg/m^3, ρ_b is the density of the air inside of the balloon in kg/m^3, V is the volume in m^3, and g is the gravitational constant of 9.81 N/kg.

You can use these equations to either determine the quantity of mass the balloon can lift, or you can re-tailor them to determine the size of the balloon required to lift a specific mass.

The body of a hot-air balloon is made up of a series of strips of fabric called *gores*. The gores are stitched together and the final shape of the balloon is revealed. Because the hot gas inside of the balloon is compressed by the surrounding cooler gases, a shape in the form of a light bulb is optimal. This shape allows for a large pocket of gas to reside at the top of the balloon and provides for a more stable design (Figure 8-30).

Figure 8-30. *The balloon launch.*

Materials

You can find the files for this project at Thingiverse.com (*http://bit.ly/14XmyAw*).

Materials List		
Item	**Quantity**	**Source**
6 × 24 in stove pipe	1	Home improvement store
8 × 24 in stove pipe	1	Home improvement store
8-32 × 1.5 in bolt	6	Home improvement store
8-32 nuts	12	Home improvement store
8-32 washers	6	Home improvement store
1/4-20 × 1/2 in bolts	6	Home improvement store
1/4-20 nuts	6	Home improvement store
1/4-20 washers	6	Home improvement store
12 in shelving brackets	3	Home improvement store
Camping stove w/ fuel	1	Sporting goods store
20 × 30 in tissue paper sheet	12	Craft store
1/8 × 32 in aluminum rod	1	Home improvement store
3/16 in heat-shrink tubing	1	Electronic supply store

Makerspace Tools and Equipment
Glue sticks
Safety glasses
Sheet metal punch
String

Making the Balloon Launcher

Step 1

Wrap a piece of string around the 8 in stove pipe and make a mark it at the overlap. Fold the string into three equal lengths and make marks that divide the string into three sections. Wrap the string back around the stove pipe and make three marks approximately 1 inch from the end of the pipe. Repeat this process for the 6 in pipe. Use a door jam as a straightedge and draw another line 1 inch in on the opposite end of the pipe. Repeat this process for each line on both pipes.

Step 2

Measure in 1 in from the end of the pipes and make a cross on your previously marked lines. This will act as a guide for your punch. Equip your sheet metal punch with a 3/16 in punch set and punch holes at each intersection. Repeat this process by aligning the angle brackets along the bottom, spaced evenly. Mark the bracket's mounting hole positions on the 8 in pipe and punch holes using a 1/4 in punch. Using the 1/4-20 nuts and bolts, mount the 3 shelving brackets to their corresponding holes on the 8 in stove pipe. Secure the brackets to the pipe by using 1/4-20 bolts, nuts, and washers.

Step 3

Position the 6 in pipe inside the 8 in pipe and line up the top and bottom holes. Pass each 6-32 bolt through the 8 in pipe hole first, add a washer and nut, and then pass the end of the screw through the hole in the 6 in pipe. This configuration allows you to suspend the hot 6 in pipe inside of the 8 in pipe thus preventing injury to the students and catching the balloon on fire (Figure 8-31).

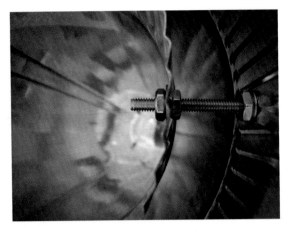

Figure 8-31. *Suspending the 6 in pipe inside of the 8 in pipe produces a heat barrier and makes for safer operation.*

complete. This offset accommodates the added material needed to glue the gores together.

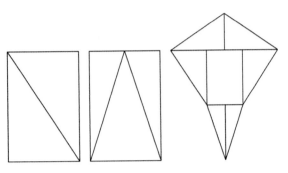

Figure 8-33. *The simplest balloon is made from single pieces of tissue paper, or those that have been cut into triangles.*

Step 4

Tighten all six fasteners to the 8 in pipe and add the remaining washers and nuts to the ends of the screws inside the 6 in pipe. Gently tighten the nuts inside the 6 in pipe until they are just tight. If you over tighten them, the walls of the pipe will distort and potentially crack. Test the stability of your launcher by placing it on a flat surface.

The base of the launcher (Figure 8-32) should be approximately 2 in above the ground and the pipes should be perpendicular with the floor.

Figure 8-32. *Launcher base.*

Making the Balloon
Step 1

Using a large format printer, or multiple sheets of paper, print out full-sized copies of the *tissueBalloon.dxf* gore template (Figure 8-33). Carefully draw a line around the perimeter of the gore template with an 0.5 in offset and cut out the template when

Step 2

Determine how many pieces of tissue paper are needed to assemble one gore and glue them together with a glue stick, end to end, and with a 0.5 in overlap. After the glue dries, neatly stack your tissue paper and place the gore template on top. Use two binder clips to secure the tissue paper and gore template in place. Slowly cut along the edge of the gore template and through all of the sheets of tissue paper. The result should be a bundle of gores with the template on top.

Step 3

One at a time, run a bead of glue down the edge of the gore and glue it to another. Repeat this process until all of the gores have been attached. Cut out a small circle of tissue paper and adhere it to the top of the balloon. This helps to prevent hot air from leaking through the opening at the top.

Step 4

Create a 9 in circle from your aluminum rod and cover the junction with a 1 in piece of heat-shrink tubing. Shrink the tubing in place with the heat gun. Attach the aluminum ring to the opening at the base of your balloon, positioning the ring inside the base and folding tissue paper around it. Glue the tissue paper in place with your glue stick. Let your balloon dry for about 30 minutes. During this time, carefully examine the seams and look for potential leaks. If you find one, smear some glue from the glue stick onto your finger and use it to close the hole.

Step 5

When the balloon is dry, it is ready for flight (Figure 8-34). With the help of your instructor, assemble the launcher and fire up the burner. Carefully position the balloon over the launcher, and as soon as the balloon is full of hot air, let it fly!

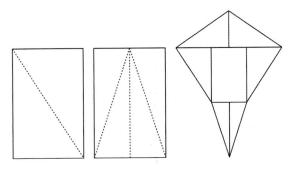

Figure 8-34. *How high can your balloon float?*

Investigation

Because hot-air balloons produce lift by enclosing a gas with less density than the surrounding air, you can calculate the quantity of lift. You can use this calculation to determine how much mass the balloon can lift given the difference in gas temperature inside and outside of the balloon.

Calculating Lift

1. Use the formulas provided at the beginning of this lab to determine how much mass your balloon should lift.

    a. How much mass can the balloon lift if the outside air temperature is 40°F and the temperature of the air inside the balloon is 100°F?

    b. How big would the balloon need to be to lift you?

Measuring Lift

1. Use a series of masses to determine how much your balloon can lift.

2. Measure the temperature outside and inside of the balloon. Run the calculations to determine how much mass your balloon should lift. Do the numbers match?

Makerspace Art

The utilization of Makerspace technology in the art classroom is a natural fit. Where technology education and engineering work to bring theoretical concepts into reality, art works to improve their design and appeal. Using these tools in art helps to bring new mediums into the classroom and exposes the students to the alternative uses of the technology:

- "Lab #10: Making a Silhouette" on page 259
- "Lab #11: Making a Mold and Casting a Part" on page 261
- "Lab #12: Making a Printing Press" on page 262

Lab #10: Making a Silhouette

Subject
    Art

Grade Level
    K–5

Lab Type
    Activity

Estimated Class Time
    45 Minutes

Objective
    Making the connection between the virtual world of computers and the real world is an important recurring concept. This lab is designed to give students the opportunity to physically create a personal silhouette

using photography, digital art, and a laser cutter.

Overview

A silhouette (Figure 8-35) is a shadow representation of an object that conveys both depth and emotion. This simple artform is commonly used to capture portraits using inexpensive materials. With the help of Makerspace technology, you can create an effective silhouette from a single digital photograph.

Figure 8-35. *Completed silhouette.*

Materials

Materials List		
Item	**Quantity**	**Source**
8.5 × 11 in black construction paper	1	Craft store
8.5 × 11 in white printer paper	1	Craft store
Glue stick	1	Craft store

Makerspace Tools and Equipment
Digital camera
Laser engraver
Photo editing software or CAD
Tripod

Setup
Step 1

Create a monotone backdrop for the student photos by using a white sheet or blank wall. Position a chair in front of the backdrop and place a strip of tape on the floor indicating the position of the student's feet. Set up the camera and tripod approximately 6 ft away from the chair and adjust the camera until it is level with the average height of a student's head.

Step 2

Configure the camera to use the countdown timer. This way, the student can take the picture while also eliminating the chance of a blurry picture due to camera movement. Separate your students into teams of two. One student will be in charge of taking the other's photo, and vice versa. Have the first student sit on the chair so that her silhouette is visible in the viewfinder or on the LCD. Let the student's partner take the picture and continue until everyone is photographed.

Be sure to use a high-contrast backdrop (Figure 8-36). This will help the students when they trace around their face.

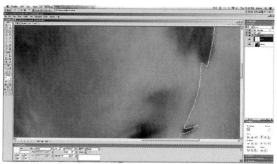

Figure 8-36. *Tracing the face.*

Procedure
Step 1

Open the *defaultTemplate.dxf* file from "Project: Creating a Default Template" on page 179, into your CAD software and import the student's photograph into the vector-drawing software of your choice. Select the Spline tool. This tool usually looks like a squiggled line. Guide the students through the process of clicking the mouse around the outline of their face. The spline tool automatically creates a curved line through the points created by the mouse clicks. If the student wants to edit his spline, he can simply move the points or nodes after he has finished the initial spline. Delete the student's photograph from the layout and position the silhouette spline into the cutting area.

Step 2

Follow your laser cutter's operating procedure for configuring the machine for cutting. Position the black construction paper on the cutting bed, send your drawing to the machine, power on the exhaust fan and air assist, and begin cutting. Hand out the silhouette cutouts to the students along with a sheet of white paper and a glue stick, and then have them glue their silhouette's to the paper.

Lab #11: Making a Mold and Casting a Part

Subject
    Art

Grade Level
    6–8

Lab Type
    Activity

Estimated Class Time
    1.5 Hours

Objective
    A significant amount of the world around us is the product of a molding process. From resins to plastics, molding makes it possible for us to quickly and inexpensively create multiple copies of an existing three-dimensional object. Understanding the fundamentals of this process helps to develop an appreciation for consumer objects and how they are made.

This project requires that safety glasses be worn throughout its entirety.

Overview

Part molding is used by industry to rapidly produce copies of an object. You can replicate this relatively simple process by using only a few materials, yielding results that are very close to perfect (Figure 8-37).

Figure 8-37. *Casting a star.*

Materials

You can find the files for this project at Thingiverse.com (*http://bit.ly/15aH2tQ*).

Materials List		
Item	**Quantity**	**Source**
Plaster of paris or equivalent	1	Home improvement store
Petroleum jelly	1	Grocery store

Makerspace Tools and Equipment
3D printer
Gloves
Masking tape
Safety glasses
Sandpaper

Procedure

Step 1

Create a small shape that you would like to cast via the solid-modeling software of your choice. Save the file and open the *moldBlankBottom.stl* and *moldBlankTop.stl* files. Create an assembly with the mold bottom and your part. Position the bottom half of your part within the mold bottom and use the Subtract tool to remove the shape's volume from the mold. Repeat the procedure for the mold top using the top half of your part. Save your work and export the files as *.stl*.

Step 2

Using a sheet of 400-grit sandpaper, gently sand flat the mold faces so that they seal tightly. Lightly coat the surface of the mold with petroleum jelly and rub clean with a paper towel. This helps the casting to separate from the mold. Align the two mold pieces together and secure with masking tape. For larger or misshapen molds, use threaded fasteners to clamp the two pieces.

Step 3

Mix up a small batch of plaster and pour it into the mold (Figure 8-38). Gently tap the mold on your work surface to eliminate as many bubbles as possible. Let the mold rest for the plaster's required set time.

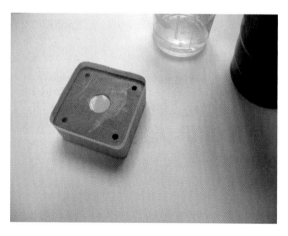

Figure 8-38. *Mix the casting material gently and pour slowly to help prevent unwanted bubbles.*

Step 4

Gently separate the two pieces of the mold and extract the part. Use a knife and sandpaper to trim the edges and finish the piece as desired.

Lab #12: Making a Printing Press

Subject
   Art

Grade Level
   9–12

Lab Type
   Learning Tool

Estimated Class Time
   1.5+ Hours

Objective
   Printing today continues to be as relevant as it was during the time of Gutenberg and his press. From books to t-shirts, printing is an artform that allows for the rapid reproduction of artwork onto almost any medium.

Overview

This lab is designed to expose students to the mechanics of paper printing through the construction of a single stamp printing press and corresponding stamp (Figure 8-39). The resulting press is capable of accepting a custom-made stamp cartridge and transferring its image to a 3 × 5 in piece of paper.

Figure 8-39. *The completed printing press.*

Materials

You can find the files for this project at Thingiverse.com (*http://bit.ly/18KHMpp*).

Materials List		
Item	**Quantity**	**Source**
1/4 in rubber sheet	1	Craft store
1/8 in plywood sheet	1	Home improvement store
1/4-20 × 1-1/4 in bolt	2	Home improvement store
1/4-20 nut	2	Home improvement store
1/4 in washer	4	Home improvement store
3 × 5 in blank cardstock	10	Craft store
Colored roll on ink	1	Craft store

Makerspace Tools and Equipment

Laser engraver

Making the Stamp

Step 1

Open the *stamp.dxf* template in your CAD software and insert the desired image or text inside the designated area. Remember to invert any text because it will print the mirror image of what you see on the screen. Some print drivers can do this for you automatically, and you can use this in lieu of inverting the text in the CAD software.

Step 2

Follow your laser cutter's operating procedure for configuring the machine for engraving and cutting stamps. Position the stamp material on the bed, send your drawing to the machine, power on the exhaust fan and air assist, and begin engraving. After engraving, the machine will cut the outline.

Making the Press

Step 1

Open the *press.dxf* file in your CAD software suite and create an array of the number of presses that you want to make. Follow your laser cutter's operating procedure for configuring the machine for cutting. Send your drawing to the machine, power on the exhaust fan and air assist, and begin cutting.

Step 2

Construct the bed by gluing the reference pins in place with a small amount of glue. When dry, assemble the press by gluing the two sides to the base. Slide on the front and back cross members to complete the primary structure. The notches are designed to lock the cross members in place when fully assembled.

Step 3

Assemble the press plate by gluing the reference pins in place and attaching the pivot mount. When dry, attach the two press arm connecting rods to the mount by sliding a washer and then a bolt through the assembly. Lightly secure the bolt with another washer and nut.

Step 4

Slide the 1/2 in dowel through one of the sides and add a wooden washer. Slide the press arm over the dowel and add another washer. Secure the dowel in place by gluing the two last wooden washers over the ends of the dowels so that it turns freely yet does not slide side to side. Attach the connecting rods to the press arm with the remaining bolt, washers, and nut.

Step 5

Assemble the stamp cartridge by gluing the reference pins in place. When dry, position the rubber stamp onto the plate and slide the assembly into the press. The reference pins should properly align the stamp to the press plate. Lower the press until it just touches the surface of the stamp. Make any adjustments to the height of the spring mounts to make the bed as level as possible.

Step 6

Raise the press and remove the cartridge. Slide a 3 × 5 in card into the press plate holder and apply ink to the stamp surface. Place the cartridge back in the press and compress to transfer the image (Figure 8-40).

Figure 8-40. *This press is great for adding graphics and text to letters and other stationary.*

Makerspace Math

The use of Makerspace technology in the math classroom opens the door to a wide range of demonstration and student interaction possibilities. You can use this technology, which is commonly thought to exist in more hands-on classrooms, to reinforce math's theoretical concepts:

- "Lab #13: Making Plane Shapes" on page 264
- "Lab #14: Making a Pythagorean Calculator" on page 265
- "Lab #15: Making a Parabolic Focuser" on page 267

Lab #13: Making Plane Shapes

Subject
Math

Grade Level
K–5

Lab Type
Learning Tool

Estimated Class Time
45 Minutes

Objective
This lab is designed to expose students to the design and creation of plane shapes by using

a simple yet effective pegboard device. In addition, students can use the device to explore the different ways in which they can create basic shapes. It also gives them the opportunity to self pace through the process.

Overview

Traditionally, the concept of plane shapes is delivered through worksheets with a series of connect-the-dot tasks that walk the student through the creation of the assigned shape, or through the use of physical shape models (Figure 8-41). Making the connection between virtual concepts and their application in a physical world is known to improve content retention in addition to making math fun.

Figure 8-41. *The completed plane shape peg board.*

Materials

You can find the files for this project at Thingiverse.com (*http://bit.ly/1asLhEV*).

Materials List		
Item	**Quantity**	**Source**
1/4 × 3 × 4 in plywood	1	Home improvement store
1/8 × 12 in wood dowel	1	Home improvement store
18 in colored yarn	1	Craft store
1 qt food storage bag	1	Grocery store

Makerspace Tools and Equipment
Laser engraver
Ruler
Sandpaper
Wire strips

Setup
Step 1

Import the *planeBoard.dxf* file into the default template's layout and create an array of boards to meet your quantity and material constraints. Follow your laser cutter's operating procedure for configuring the machine for cutting. Send your drawing to the machine, power on the exhaust fan and air assist, and begin cutting.

Step 2

Each pegboard requires 10 1-inch pegs. Mark each wooden dowel with 1 in graduations using a pen and use the largest gauge cutter on your wire strippers to cut off each peg. Feed the yarn through the mounting holes located above the peg holders and secure it in place with a knot. Store the board and the pegs in a ziplock bag to prevent the loss of parts.

Check that the ends of the pegs are free of splinters by gently dragging the ends across the sheet of sandpaper until smooth (Figure 8-42).

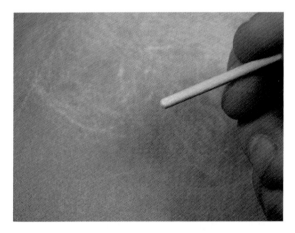

Figure 8-42. *Smoothed dowel.*

Procedure
Step 1

Set up the board by unwrapping the yarn and laying it flat on your work surface. Place the pegs into their respective positions at the top of the board. Use the pegs to create as many plane shapes as possible (Figure 8-43).

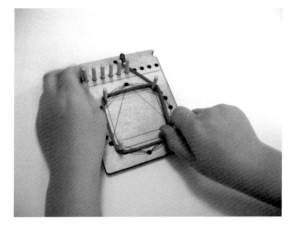

Figure 8-43. *How many shapes can you create?*

Lab #14: Making a Pythagorean Calculator

Subject
Math

Grade Level
6–8

Lab Type
Learning Tool

Estimated Class Time
45 Minutes

Objective

As students become more and more reliant on calculators, their ability to predict the numerical outcome of a problem and know when the calculated outcome is correct is greatly diminished. Providing students with tools that provide a physical representation of their calculation helps develop better math skills and gives them the ability to more accurately predict outcomes prior to using a calculator.

Overview

The Pythagorean theorem is a widely used formula for calculating the length of any side of a right triangle when two lengths are known:

$$A^2 + B^2 = C^2$$

This formula is widely used in engineering for determining material lengths and can easily be calculated without a conventional calculator. This lab is designed to produce a measurement tool that is capable of calculating and demonstrating the correlation between all three sides of a right triangle. The resulting tool (Figure 8-44) serves as an excellent means for students to make connections between their formulas and the physical world.

Figure 8-44. *The completed calculator.*

Materials

You can find the files for this project at Thingiverse.com (*http://bit.ly/184jjcg*)).

Materials List		
Item	**Quantity**	**Source**
1/8 in plywood sheet	1	Home improvement store
4-40 × 1 in pan head machine screw	2	Home improvement store
4-40 wing nut	2	Home improvement store
#4 washer	1	Home improvement store

Makerspace Tools and Equipment
Laser engraver

Setup
Step 1

Open the *pyCalc.dxf* file in your CAD software suite and create an array for the quantity of calculators that you want to make. Follow your laser cutter's operating procedure for configuring the machine for cutting.

Position the sheet of plywood on the cutting bed, send your drawing to the machine, power on the exhaust fan and air assist, and begin cutting.

Procedure
Step 1

Place one of the large wooden washers over a screw and then add a position indicator. Repeat this process for the remaining indicator. These indicators are used to identify the length of all three sides.

Step 2

Attach the hypotenuse gauge to the A^2 / B^2 side by gently sliding it over the two previously made indicators. They should move freely in the A^2 / B^2 gauge channels. Place another large wooden washer and then a steel washer over the screw and secure it in place by using a wing nut.

Step 3

Operate the gauge by sliding either the A^2 or B^2 sides into their desired position. You can read the angle of the B^2 side relative to the hypotenuse by using the angle gauge.

You can read the measurement (Figure 8-45) by noting the length of all three sides using the indicators.

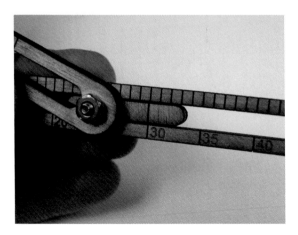

Figure 8-45. *Using the calculator.*

Lab #15: Making a Parabolic Focuser

Subject
  Math

Grade Level
  9–12

Lab Type
  Learning Tool

Estimated Time
  1.5 Hours

Objective
  Math lessons typically utilize virtual concepts with pen and paper explanations. The use of Makerspace technology to bring physical representations of the mathematical theory can help solidify student comprehension. The task of calculating and manipulating parabolic shapes and determining their focal point lends itself perfectly to such technology.

Overview

Algebra teaches about the fundamentals for calculating the shape of a parabola. Because a parabola has a definite focal point, it is well suited for focusing waves. This lab utilizes both 3D CAD software and a 3D printer to construct a tool (Figure 8-46) that demonstrates the location of the focal point of a calculated parabola.

Materials

You can find the files for this project at Thingiverse.com (*http://bit.ly/184jkwM*).

Materials List		
Item	**Quantity**	**Source**
1/8 × 4 in wooden dowel	1	Craft store
Aluminum foil	1	Grocery store

Makerspace Tools and Equipment
3D printer
Spray adhesive

Figure 8-46. *The focuser.*

Setup Procedure

Step 1

  Create a table of 10 coordinates, in addition to (0,0), that illustrate the $0.1X^2$ parabola. Open your 3D CAD software and create a new sketch on the Front plane, and set your measurement units to millimeters. Select the Point tool and produce 11 randomly placed points. One at a time, select a point and assign it the coordinates from your table. Use the Spline tool (Figure 8-47) to select the points from left to right. Exit the tool and you should have outlined the shape of your parabola.

Figure 8-47. *The spline tool produces a Bezier curve between two points and works well for reproducing parabolas.*

Step 2

Select the Offset tool and produce an offset line 3 mm to the outside of your parabola and complete the shape by using the Line tool. Use the Line tool to bisect your parabolic shape and then use the Trim tool to trim away half of the outline.

Step 3

Select and use 3D Revolve and revolve your outline, turning your line drawing into a three-dimensional model. Make a new sketch on the Top plane and create a circle with a 20 mm diameter at the origin. Extrude the circle 3 mm above the Top plane.

Step 4

Create a new sketch on the Top plane and make another concentric circle at the origin with a 1/8 in diameter. This circle will be used to cut a hole for the dowel. Select the Extrude Cut tool and extrude this circle 1 cm through the model. For the 3D printer to construct your model it needs to be saved in the proper format. Select File, click Save As, and then click STL as the file type. Open the STL in your 3D printer's control software. This software will "slice" your model into multiple layers and convert it to code that controls the printer. Generate the code and print.

Step 5

Carefully remove your parabolic model from the printer and clean away any excess plastic. Check and make sure that the dowel passes through the 1/8 in hole. Cut a small piece of aluminum foil from the roll and lay it on your desk, shiny side up. Spray a small amount of adhesive into the model and gently press in the aluminum foil. Make sure you smooth it

as much as possible (Figure 8-48) because wrinkles will distort the reflection.

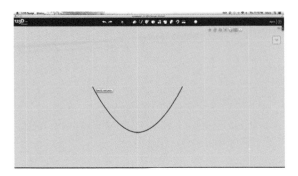

Figure 8-48. *Rub the foil with a paper towel to polish and smooth the surface.*

Step 6

Push the dowel into the parabolic model and make a mark with your pen indicating the base of the parabola (Figure 8-49). Position your model in front of the light source and mark the focal point with your pen.

Figure 8-49. *The focuser.*

Index

Symbols

We'd like to hear your suggestions for improving our indexes. Send email to index@oreilly.com.

C

E

Eagle library components, 128–129
- device, 129
- package, 128
- symbol, 129

EAGLE PCB (CadSoft), 127–139, 147
- board design, 137–139
- Board Layout, Creating project, 138–139
- circuit boards, generating CAM files for, 147–147
- Creating a Schematic project, 132–136
- custom libraries, creating, 129–132
- free alternatives to, 127
- libraries, 128–132
- schematic layouts in, 132–136
- schematics, creating, 134–136

electric hand shears, 2
electric motors, 249–251
electrical characteristics (datasheet section), 127
electrical tape, 73
electronics, 125, 169
- Arduino Development Platform, 161
- board design, 137–139
- Board Layout, Creating in EAGLE project, 138–139
- circuit boards, making, 147–150
- computer-based circuit design, 127–139
- datasheets, 126–127
- design process for, 126
- hardware, interfacing with, 158–169
- microcontrollers, 150
- parts, sources for, 126
- resources, 126
- schematic layouts, creating, 132–136
- serial communications, 152–158

- soft circuits, 140–147
- Static Sticky project, 145–147
- Toner Transfer Circuit Board project, 148–150

electronics workbench, 12
elementary school projects
- Lemon Battery, 235–237
- Plane Shapes project, 264–265
- Sail Car project, 252–253
- Silhouette project, 259–261
- Water Rocket and Launcher, 243–246

emails vs. phone calls, 11, 10
endstops (CNC), 217–218
engrave direction (engraving setting), 185
engraveable materials, 190
engraving, 190–192
engraving settings, 185
- direction, 185
- image dithering, 185
- power, 185
- resolution, 185
- speed, 185

epoxy glue, 76
- using, 77

equipment
- checklist for, 14
- common, 2
- common large, 2
- determining, 1
- noise comparison, 5
- quantity, 2
- through donation, 10
- types, 2

equipment types, 2
- clamps, 69–70
- drill bits, 53–55
- hammers, 39–41
- saws, 44–45
- screwdrivers, 30–31
- vises, 71–71
- wrenches, 37–39

ERC (schematic layout tool), 133
exhaust systems, alternatives to, 189
extrude cut (CAD), 200
extruder (3D printers), 221–226
- drive mechanism of, 221

- feed mechanism, 221–222
- heated-barrel assembly, 223–226

extrusion tool (CAD), 200

F

fasteners, 78–84
- common features, 78
- diameter, 79
- grommets, 79–80
- head styles, 78
- nonthreaded mechanical, 79
- nuts, 82–83
- rivets, 79–80
- screws, 80–82
- thread sizes, 78
- washers, 83

features (datasheet section), 127
feed mechanism (3D printers), 221–222
feedrate parameter (G-Code), 212
field magnet, 249
filament
- actively cooling, 228
- heater-barrel assembly and, 223
- temperature, monitoring, 226

Filament Spool project, 202–206
files, 46–47
- coarseness, 46
- common features, 46
- grade, 46
- half-round, 46
- rectangular, 46
- round, 46
- using, 47

fill density parameter (G-Code), 212
fill pattern parameter (G-Code), 212
fillet edges in 3D design, 199
fillet tool (CAD), 201
finger joints, 176
- variations, 177

finishing stitches, 145
firecode, 3

N

names (schematic layout tool), 133
National Electrical Manufacturers Association (NEMA), 207
National Fire Protection Association, 3
needle-nose pliers, 33
NEMA 5-15 outlet, 3
NEMA 5-20 outlet, 3
nested fastener joint, 178
Newton, Isaac, 243
nichrome heater, 223
    nozzle design for, 226
    soldering, 224
no-clean fluxes, 101
noise
    location choice and, 5
    power tools and, 5
nonmarking pliers, 33
nonmarring hammers, 40
nonprofit organization, 9
nonvolatile memory, 151
nozzle (3D printers), 226
nozzles, printing, 244
nuts, 82–83
    acorn, 82
    captive, 82
    hex, 82
    lock, 83
    wing, 83

O

Occupational Safety & Health Administration (OSHA), 3
Odyssey of the Mind (website), 9
open-ended wrench, 37
optical switch, 218
outlets, 3
    NEMA 5-15, 3
    NEMA 5-20, 3
    risks, 3
    timers, 12
    types, 3
outreach tools, 6

overhangs and 3D printing, 200
overheated joints, 103, 110

P

packaging information (data-sheet section), 127
painted brass, 191
painters tape, 3D printer build platforms and, 229
parabolas, calculating shapes of, 267
Parabolic Focuser project, 267–268
parts files (CAD), 198
patterns tool (CAD), 201
PC (Polycarbonate), 227
    build platforms for, 229
PEEK, 223
permanent magnet rotor, 250
PET (polyester) film tape, 230
Phillips screwdrivers, 30
phone calls
    emails vs., 10
phone calls vs. emails, 11
photoelectric effect, 238
photovoltaic cells, 238
pin definition (datasheet section), 127
pipe wrench, 38
pitch, thread cutters, 42
pivot, pliers, 32
PLA (Polylactic acid), 227
Plane Shapes project, 264–265
plasma cutter, 2
plastic drill bits, 54
plastic, build platforms for, 229
plastic, drilling, 55
plastics, engraving, 191
plated vias, 137
pliers, 31–34
    combination-jaw, 32
    general tool use, 33
    gripping, 32–34
    grips, 32
    head, 32
    locking, 32
    needle-nose, 33

nonmarking, 33
    pivot, 32
    tongue-and-groove, 33
    wrenches and, 36
pneumatic hand shears, 2
pneumatic punch, 2
point diameter of punches, 49
points, of drills, 52
Polygon tool (board design for electronics), 137
polyimide film tape, 74, 230
poor solders, identifying
    surface-mount soldering, 109–110
    through-hole soldering, 103–105
pop rivets, 79
portable power tools, 55–63
    battery chemistry comparison, 56
    body, 56
    circular saws, 56
    common features, 56
    drills, 57
    hammer drills, 58
    head, 56
    impact drivers, 58
    jig saws, 59
    power sources, 56
    reciprocating saws, 60
    sanders, 60
    scroll saws, 59
    sewing machines, 61
potentiometer, 215
power
    Arduino considerations, 163
    calculating consumption, 3
    cut settings, 186
    engraving settings, 185
    for 3D printers, 197
    location choice and, 3
    microcontrollers, 151
    real, 4
    total, 4
power consumption
    calculating, 3
    display, 3
    measure, 3
power factor, 4